Make:

FUSION 360 for MAKERS

Design Your Own Digital Models for 3D Printing and CNC Fabrication

Lydia Sloan Cline

Maker Media, Inc.
San Francisco

Printed in Canada

Maker Media, Inc.
1700 Montgomery Street, Suite 240
San Francisco, CA 94111

Maker Media books may be purchased for educational, business, or sales promotional use. Online editions are also available for most titles (safaribooksonline.com). For more information, contact our corporate/institutional sales department: 800-998-9938 or corporate@oreilly.com.

Editorial Director: Roger Stewart
Editor: Patrick DiJusto
Copy Editor: Elizabeth Welch, Happenstance Type-O-Rama
Technical Editor: Paul Yanzick
Proofreader: Scout Festa, Happenstance Type-O-Rama
Interior and Cover Designer: Maureen Forys,
Happenstance Type-O-Rama
Compositor: Kate Kaminksi and Maureen Forys,
Happenstance Type-O-Rama
Indexer: Valerie Perry, Happenstance Type-O-Rama

May 2018: First Edition

Revision History for the First Edition

2018-05-15 First Release

See oreilly.com/catalog/errata.csp?isbn= 9781680453553 for release details.

978-1-68045-355-3

Safari® Books Online

Safari Books Online is an on-demand digital library that delivers expert content in both book and video form from the world's leading authors in technology and business. Technology professionals, software developers, web designers, and business and creative professionals use Safari Books Online as their primary resource for research, problem solving, learning, and certification training. Safari Books Online offers a range of plans and pricing for enterprise, government, education, and individuals. Members have access to thousands of books, training videos, and prepublication manuscripts in one fully searchable database from publishers like O'Reilly Media, Prentice Hall Professional, Addison-Wesley Professional, Microsoft Press, Sams, Que, Peachpit Press, Focal Press, Cisco Press, John Wiley & Sons, Syngress, Morgan Kaufmann, IBM Redbooks, Packt, Adobe Press, FT Press, Apress, Manning, New Riders, McGraw-Hill, Jones & Bartlett, Course Technology, and hundreds more. For more information about Safari Books Online, please visit us online.

How to Contact Us

Please address comments and questions to the publisher:

Maker Media, Inc.
1700 Montgomery Street, Suite 240
San Francisco, CA 94111

You can send comments and questions to us by email at books@makermedia.com.

Maker Media unites, inspires, informs, and entertains a growing community of resourceful people who undertake amazing projects in their backyards, basements, and garages. Maker Media celebrates your right to tweak, hack, and bend any Technology to your will. The Maker Media audience continues to be a growing culture and community that believes in bettering ourselves, our environment, our educational system—our entire world. This is much more than an audience, it's a worldwide movement that Maker Media is leading. We call it the Maker Movement.

To learn more about Make: visit us at makezine.com. You can learn more about the company at the following websites:

Maker Media: makermedia.com
Maker Faire: makerfaire.com
Maker Shed: makershed.com
Maker Share: makershare.com

Contents

PART II   Make Some Stuff

Preface

So you have a 3D printer and want to join the new Industrial Revolution! You want to make original designs and maybe even customize those designs to specific people. That means you can't just download and print other people's files—you need to rock that software yourself. Well, if you are just learning how to do that or have outgrown the "starter" apps, you've come to the right place! With this book you'll soon be making your own designs with Autodesk's Fusion 360 software. So pour your favorite beverage, pull up a computer for playing along, and read on.

What Is Fusion 360?

Fusion 360 is a 3D design program (Figure P-1). Autodesk chose that name because many disparate software types are rolled into it. You can do computer-aided design (CAD), computer-aided manufacturing (CAM),

FIGURE P-1: Fusion 360

and computer-aided engineering (CAE) tasks. Its CAD capabilities let you model just about whatever you want with it (Figure P-2). Its CAM capabilities let you generate NC code for CNC cutting machines—up to five axes—so you can cut, carve, and do rotary work. Its CAE capabilities let you analyze the model for optimal performance. Fusion 360 is not a building information modeling (BIM) program, but you can export its models in a format that will import into a BIM program.

Fusion 360 can animate (make videos), render (add lifelike colors and textures), and display the model as a set of scaled, 2D drawings. It lets you arrange the model's parts into an assembly. Many proprietary file types can be imported into it. You can also collaborate on your projects using A360, an online Autodesk workspace integrated with Fusion.

Fusion 360 is a robust program, and it can be complicated, but it is still suitable for all levels of user. Whether you're new to modeling, a hobbyist, or an experienced engineer, you'll find this software useful. Browse the Fusion 360 gallery to see designs created by users like you (**gallery.autodesk.com/fusion360**). Many are downloadable.

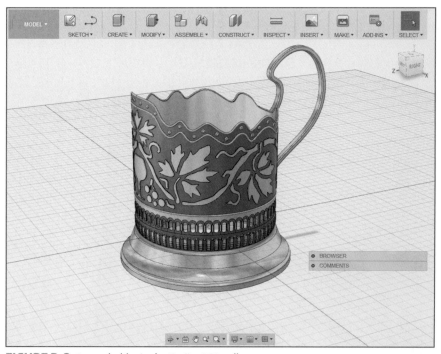

FIGURE P-2: A mug holder in the Fusion 360 gallery

In the Cloud

Fusion operates online, and all your files are saved online. This means you'll never lose them and always have access to them. However, you download the program onto your PC, Mac, or mobile device. Working online offers the most functionality, but if your Internet isn't working, or if Autodesk's servers are down, you can work offline. You can also store your files offline.

Types of Modeling

Modeling is 3D drawing. Fusion lets you model three ways: freeform, solid, and surface.

FREEFORM MODELING This makes hollow models composed of polygons (Figure P-3). You sculpt the model's surface by pushing and pulling the polygons to make curved, flowing shapes.

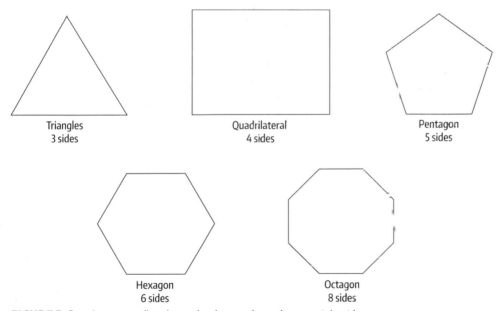

Triangles
3 sides

Quadrilateral
4 sides

Pentagon
5 sides

Hexagon
6 sides

Octagon
8 sides

FIGURE P-3: Polygons are flat planes that have at least three straight sides.

There are different kinds of freeform modeling, such as polygonal, NURBS, and T-splines. Fusion's freeform modeling uses T-splines, a system that combines the best of polygonal and NURBS. It lets you add detail only where needed, create nonrectangular polygons, and easily edit complex freeform models (Figure P-4). T-spline surfaces can contain areas with differing levels of detail, and control (editing) points can be added only where needed. They can be converted to a NURBS surface if you want to export your model into another program. You can also convert it to a solid model.

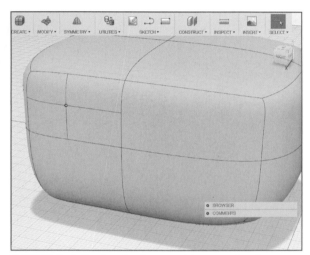

FIGURE P-4: T-spline freeform models are easily edited.

SOLID MODELING In this system you add and subtract primitives—geometric forms such as cubes and cylinders—to and from each other. This process is technically called constructive solid geometry (CSG) modeling. You can also sketch 2D shapes and perform operations to turn them into 3D volumes, a process technically called boundary representation (BREP) modeling (Figure P-5). When sliced open, solid models have volume inside.

SURFACE MODELING This makes solid-appearing objects, but when sliced open, they're hollow inside. They only have surface area, not volume.

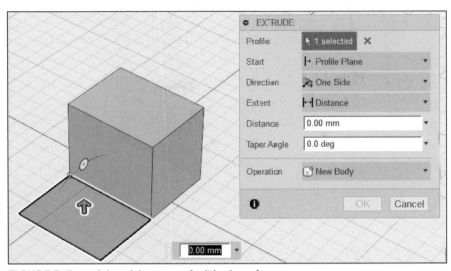

FIGURE P-5: A solid model consists of solid volume forms

Parametric and Direct Modeling

Fusion lets you model two ways: parametrically and directly.

▶ Parametric: You design using constraints and parameters. You can edit dimensions individually; for example, if you model a wall 6 inches wide, that wall will remain 6 inches wide even if you double the overall floor plan. Parameters can be changed by accessing them on a timeline of actions that appears as you work. Everything upstream will automatically update to accommodate the change.

▶ Direct: You manipulate a model by simply pushing, pulling, and twisting it. It's easier to edit, but the lack of a timeline means you can't change features and see everything upstream automatically update in response to those changes.

Generally speaking, direct modeling is easier and more intuitive than parametric modeling. If you're already used to working with a direct modeling CAD program, you might choose Fusion's direct modeling mode to most resemble what you're used to. Parametric design requires a different thought process because of feature dependencies—that is, almost every item's existence depends on something else existing. However, parametric modeling is best for complex engineered items because the ability to make size changes and have everything update around those changes can be invaluable during the design process.

Some parametric modeling functions are unavailable in direct modeling, and some direct modeling functions are not available in parametric. Solid and surface models can be created parametrically or directly. Freeform models can only be created directly.

What We'll Do in This Book

This book shows beginning and intermediate Makers how to use the Fusion 360 software. The goal is to build confidence with projects of increasing complexity, each showing new techniques. No prior modeling knowledge with this, or any CAD program, is assumed. We'll do parametric, direct, solid, surface, and freeform modeling, and make an assembly with joints. We'll discuss how to install Netfabb for Fusion, an add-on that analyzes digital models for problems. We'll run one project through Slicer, which is another Fusion add-on that makes 2D pattern pieces from the model. We'll make 3D documentation drawings from a project, and run another project through the CAM workspace.

We'll design projects from scratch, make them from imported sketches, and edit downloaded files. We'll also discuss best practices for optimizing a digital model for fabrication.

The first four chapters contain general sketching, modeling, and editing information, so read them first. The projects reiterate that information, explain it further, and build on it. Save the files after completion, because some projects are used in later chapters. Sidebar tips are included as appropriate—tips on using the software, printing, and solving common problems.

What's Needed to Run Fusion

These are the system requirements currently needed to run the software:

Operating system	Apple macOS Sierra v10.12; macOS X v10.11.x (El Capitan); macOS v10.10.x (Yosemite); Microsoft Windows 7 SP1, Windows 8.1, or Windows 10 (64-bit only)
CPU type	64-bit processor (32-bit not supported)
Memory	3 GB of RAM (4 GB or more recommended)
Graphics card	512 MB of GDDR RAM or more, except Intel GMA X3100 cards
Disk space	~2.5 GB
Pointing device	Microsoft-compliant mouse, Apple Mouse, Magic Mouse, MacBook Pro trackpad
Internet	A DSL Internet connection or faster

Note that while Fusion can be used with the pointing devices shown in the chart, a mouse that has two buttons and a scroll wheel is best. The left button selects, the right button brings up context menus and serves as the Enter key, and the scroll wheel rotates and zooms.

3D Mouse Option

An option for both PC and Mac users is 3Dconnexion's Space Navigator (Figure P-6), designed for modeling software. It works like a game joystick, combining a modeling program's zoom, pan, and orbit navigation tools into one. It also tilts, spins, and rolls. The Space Navigator is used along with the traditional mouse. You left- and right-click with the traditional mouse, and move around the model with the 3D one. You can also program the tools you use the most into it, enabling you to access them by clicking anywhere on the screen.

FIGURE P-6: The Space Navigator mouse is designed for 3D modeling.

Download Fusion 360

You need a free Autodesk account to start your Fusion adventure. Create yours here: **http://accounts.autodesk.com/register**. This account will work on all Autodesk websites. Then download Fusion at **http://autodesk.com/products/fusion-360/subscribe**. It's free for personal use and small businesses. All others purchase a monthly or annual subscription, with a Standard or Ultimate option (the latter offers advanced simulation and manufacturing capabilities).

To get a free version, click the Free Trial tab at the site. On the next page, choose "Are you a student?" or "Are you a startup or hobbyist?"

1

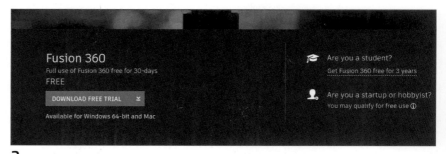

2

FIGURE P-7: How to download a free version for students and hobbyists

Definitions

These terms are used a lot in subsequent chapters, so here's an all-in-one-place reference for them.

ASSEMBLY A multipart design held together with joints to provide motion.

BODY A single 3D form.

CAD Computer-aided design.

CAM Computer-aided machinery.

CNC Computer numerical control. The automation of machine tools by computer.

COMPONENT An invisible container that holds bodies, other components, sketches, and any other construction components. Also an independent part of an assembly.

CONSTRAINTS Restrictions applied to sketches.

CONSTRUCTION GEOMETRY Planes, axes, and points. Also called construction objects.

DXF Drawing Exchange Format file developed by Autodesk. It's a universal format for storing CAD models.

FACE A flat surface bordered by edges.

FDM Fused deposition modeling. Also called fused filament fabrication (FFF). An additive manufacturing process that uses an extruder to melt plastic filament and deposit it onto a build plate in the form of the digital model.

FEATURE A specific item whose characteristics you create. Features can also be construction operations. For example, holes, fillets, and chamfers are features, as are the results of extrude, revolve, mirror, circular, and rectangular pattern operations.

FEATURE DEPENDENCY When a feature is dependent on another to exist. For example, a chamfer needs an edge.

FILAMENT Plastic thread that is wound onto spools.

GEOMETRY An umbrella term for sketches, points, edges, 3D forms, and all other construction objects. In a CAD program, geometry is technically the mathematical description of a shape.

JOINT A point at which parts are joined to enable motion. In an assembly of parts, joints let you create animations to show the movement.

G-CODE Common name for the most widely used numerical control (NC) programming language.

MESH MODEL A design made of polygons and hollow inside (think air-filled balloon).

MILLING A cutting process that removes material from the surface of a piece of stock in a series of passes.

MODEL The digital design. This term usually refers to the whole design, as opposed to *feature,* which is an item on the design.

MODELING The process of working inside 3D software to make a digital design.

NC Numerical control. A programming language that a computer-controlled machine can read.

ORTHOGRAPHIC DRAWINGS 2D drawings that describe a 3D object.

ORTHOGRAPHIC VIEW A 2D view of a 3D object, such as from the top or left side.

PARAMETER An equation that sets conditions. For example, to make one hole consistently half the size of another, tie the two together with a parameter equation.

PART A body or component. You can make them yourself as well as use parts made by others.

PLA A corn-based plastic filament whose ease of working with makes it suitable for beginners. It emanates mild fumes, doesn't warp much, and is hard, brittle, and shiny.

POLYGON Closed, flat plane with at least three straight sides. Types of polygons are *triangles* (3 sides), *quads* (4 sides), and *ngons* (more than 4 sides).

ROOT A file's default component. Its title is shown in the browser's top field.

SLICER A software program that converts an STL file into language a 3D printer can read. It slices a digital file into cross sections that become the print's physical layers. Popular slicers are MakerBot Desktop, Simplify3D, and Cura.

SOLID MODEL A design with continuous volume (think rock).

STL A file format for 3D printing. It is a shell that contains a mesh model.

SUBASSEMBLY One component.

TOPOLOGY The structure or skeleton of a model.

VERTEX (VERTICES, PLURAL) An angular point of a polygon.

Now join me in Chapter 1 and let's get started!

Additional Resources

3D Hubs Talk (**www.3dhubs.com/talk**): Get answers to your 3D printing questions.

Autodesk Fusion 360 YouTube Channel (**www.youtube.com/user/AutodeskFusion360**): Lots of instructional videos.

Fusion 360 Gallery (**gallery.autodesk.com/fusion360**): See user projects here.

Fusion 360 Forums (**forums.autodesk.com/t5/fusion-360/ct-p/1234**): Get answers to your questions here.

Lydia's YouTube Channel (**youtube.com/profdrafting**): Has lots of Fusion and other software tutorials.

Part I

UNDERSTANDING FUSION 360

THE FUSION 360 INTERFACE

In this chapter we'll explore Fusion's interface, navigate it, make some sketches and solids, save and upload a file, and discuss where files are stored.

We downloaded Fusion in the preface. On a Mac, launch it from the Applications folder; on a PC, click its desktop icon. A window will appear, asking you to sign in (Figure 1-1). If you haven't made an Autodesk account yet, go to **accounts.autodesk.com** and do so.

 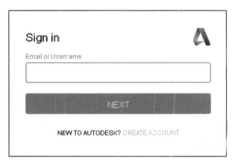

FIGURE 1-1: The Fusion desktop icon and sign-in screen

THE INTERFACE

Figure 1-2 shows the interface. It consists of a Data panel, gridded workplane, Browser panel, toolbars, ViewCube, and axis lines that intersect at a point called the origin. Let's look at them.

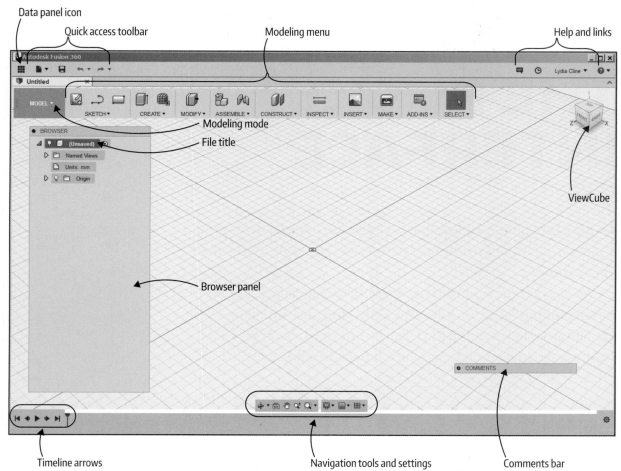

Data panel icon

Quick access toolbar

Modeling menu

Help and links

Modeling mode

File title

ViewCube

Browser panel

Timeline arrows

Navigation tools and settings

Comments bar

FIGURE 1-2: Fusion 360's interface

Data Panel

Click the grid icon to open the Data panel, where you can access your existing designs, import files, add collaborators, and see who currently has a file open. Double-clicking a file's thumbnail opens it. Dragging the thumbnail into the open workspace brings those components into the open file (you must save the open file before you can drag anything into it). Click the grid icon again to close.

Quick Access Toolbar

This toolbar shows File and Save icons and the undo and redo arrows. Click the File icon to access Fusion functions such as New Design and Export (Figure 1-3).

Help and Links

Click the dropdown arrows to access links to forums, tutorials, a gallery of user-generated files, and more (Figure 1-4). The arrow next to your name accesses the Preferences window, which offers user interface (UI) customization options (Figure 1-5). For example, if you don't like tooltips appearing, want to reverse the zoom direction, or change the default modeling orientation, you can do so. Settings are stored in the cloud, so they'll remain in place if you work on different computers.

FIGURE 1-3: Click the File icon to access Fusion functions.

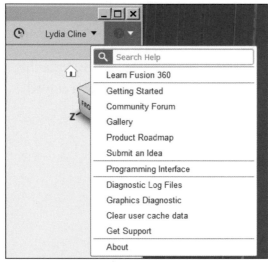

FIGURE 1-4: Access links for community help and tutorials here.

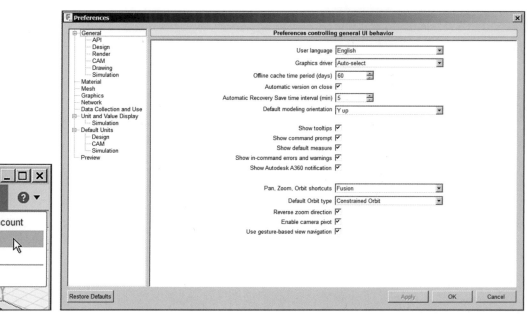

FIGURE 1-5: Customize the UI in the Preferences menu.

Modeling Toolbar

This toolbar contains the modeling tools. Click the dropdown arrow next to each menu entry to see the submenus. You can also type **S** anywhere on the screen to make a search box appear; type the tool or its hotkey into that box to activate the tool. See Additional Resources (at the end of this chapter) for the URL to the Fusion Shortcuts and Hotkey Guide.

Let's use a tool now. Click the Sketch menu's dropdown arrow, and then click Rectangle and 2-point Rectangle (Figure 1-6). Note that when you hover over a tool that isn't already in the toolbar, a curved arrow appears that you can click to add it to the toolbar.

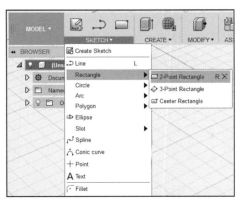
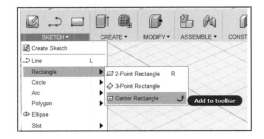

FIGURE 1-6: The 2-point Rectangle (left). If a tool isn't already in the toolbar, a curved arrow will appear when you hover over it; click to add (right).

Three planes appear; one is horizontal and two are vertical. Click the horizontal one and then click two points of the rectangle (Figure 1-7). Click Stop Sketch to finish (Figure 1-8).

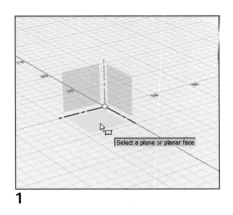

1

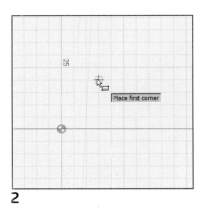

2

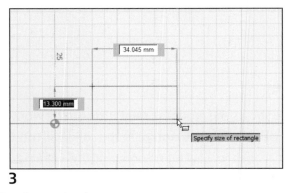

3

FIGURE 1-7: Drawing a rectangle

Right-click the rectangle to select it and open a context menu. Choose Press Pull from that menu and drag the arrow up. Press Enter and you'll have made a solid form called a *body* (Figure 1-9).

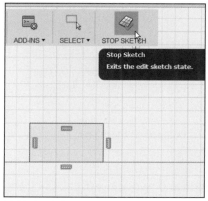

FIGURE 1-8: Click Stop Sketch to finish the rectangle and return to a 3D view.

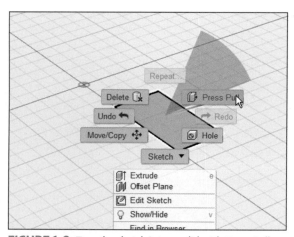

FIGURE 1-9: Turn the sketch into a solid with Press Pull.

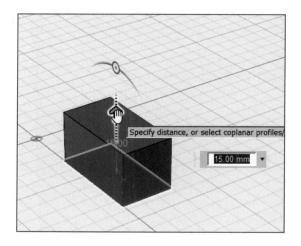

ViewCube

The ViewCube shows the model's orientation on the work plane. Left-click the cube and drag to rotate it; the solid will rotate with it. Click the cube's sides to view it orthographically—that is, as a top, front, or side view. Hover the mouse over the cube to make a house icon and a dropdown arrow appear. Clicking the house icon returns the sketch plane to its default position and zooms in to fit the model on the screen; clicking the dropdown arrow reveals options for viewing the model in perspective, orthographically, or in perspective with orthographic faces (Figure 1-10).

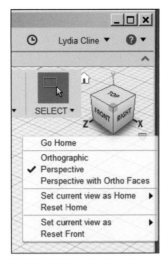

TIP Because orthographic views lack the line convergence of perspective, they're better for aligning edges. A model set to display orthographically will remain that way until changed back.

Modeling Workspaces

Click here to set the workspace's modeling mode (Figure 1-11). Switching modes also switches the toolbar to show the relevant tools. The workspaces are as follows:

- **Model:** Create and edit solid models.
- **Sculpt:** Create and edit T-spline models (organic forms).
- **Patch:** Create and edit surface models.
- **Sheet Metal:** Create and edit designs made of thin sheets.
- **Render:** Apply realistic colors and textures to the model. It's only a display and has nothing to do with a physical 3D print. You can render locally or in the cloud. Cloud rendering is much faster, but it requires purchased credits.
- **Animation:** Create videos showing how the design should work.
- **Simulation:** Run tests to see how the model will perform under loads and stress.
- **CAM:** Generate machine toolpaths to cut the model out on flat pieces of material.
- **Drawing:** Generate a multiview layout of the model.
- **Mesh:** Modify and repair STL files. This workspace appears when an STL file is inserted.

FIGURE 1-10: The ViewCube shows the model's orientation on the work plane.

FIGURE 1-11: The Fusion modeling modes

Sculpt is actually a sub-workspace, accessed through the Model and Patch workspaces. The Sketch menu is a sub-workspace, too, and both the Model and Patch workspaces have it. The Sheet Metal workspace is available only when the timeline (discussed shortly) is enabled.

Browser Panel

The Browser panel lists document settings, views, and the origin, axes, and planes. Click Document Settings to access units; you can change them here. Click Named Views to see the model in a top, front, or side position. Click the origin to see the different planes, useful for choosing a sketch orientation.

Every Fusion document has a single, default component called the *root* component. The browser's top field shows its name, which is Unsaved until you save it. Right-click it to access functions. As the model evolves, everything—sketches, bodies, components, and assemblies—is listed in the Browser panel. You can edit these items through the browser and control their visibility. Items appear with default names like *Component 1:1*, *Component 1:2*, or *Body*. The first number is the version number, and the number after the colon is the copy number. You can rename all browser entries, a good practice for telling them apart as their numbers grow. Click the text field to activate it, type a new name, and then press the Enter key to finish.

All panels can be undocked by dragging their top bar. Re-dock by dragging the panel over a workspace edge until a vertical green line appears (Figure 1-12) and then releasing the mouse. Keep the browser open or click the minus sign (–) to minimize it.

Selecting

The first entry under the Select menu is the Select tool. Clicking it deselects whatever tool you're in. Dragging it from the upper left to the lower right makes a selection window that selects everything inside it. Dragging it from the lower right to the upper left makes a crossing window, which selects everything it touches.

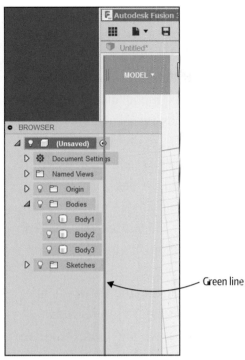

FIGURE 1-12: To dock or undock the browser, drag it over a workspace edge until a green line appears.

You can also select bodies and components by clicking their browser entry. This ensures that you select the whole item. When you select an item by clicking it directly or by dragging a window around it, sometimes you might not select the whole item, which will result in tools not working on it.

Each browser entry has a light bulb in front of it. When the bulb is yellow, the item is visible; when the bulb is gray, the item is hidden. When there are subcategories, such as the category Body and five bodies under it, the category can be turned off but an individual item under it turned on. Likewise, a category can be turned on but an individual item under it turned off.

Timeline

Also called the *history tree*, the timeline is a row of icons that appears at the bottom of the screen while you're designing parametrically. There's an icon for each action performed (Figure 1-13). The timeline grows with the project, and multiple icons can be grouped (click the first icon, hold the Shift key down, and click the last icon) to make it more manageable.

Double-click an icon to select its corresponding feature in the model; right-click an icon to access a context menu; drag an icon left or right to change the order in which operations are calculated. Move the slider to the left to "go back in time"—that is, to revisit earlier operations. A yellow icon means that feature has a warning; perhaps you've deleted something that Fusion needs to keep the design intact, but Fusion cached it, enabling you to continue working. A red icon means there's an error; an example is the edge that a fillet used was deleted, meaning the fillet can't be constructed anymore. It's best to fix warnings and errors as they arise.

You can turn off the timeline by right-clicking the browser's title field and clicking Do Not Capture Design History. This puts you in direct modeling mode, which you may find easier and more intuitive than parametric modeling mode. To turn the timeline back on, right-click the browser's title field and click Capture Design History. The timeline will start from that point; you permanently erase all previous timeline icons when you enter Direct Mode.

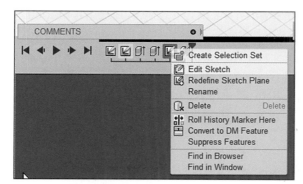

FIGURE 1-13: The timeline has icons of all operations performed. (Top) Red and yellow highlights indicate warnings or problems. (Bottom) Right-click an icon for a context menu.

Navigation Panel

The Navigation panel contains tools that let you zoom, pan, and orbit. It also has display settings that affect the workspace and how the model appears (Figure 1-14). For instance, you can change the grid size, change how the mouse snaps to the grid, see the model as wireframe, and see multiple views of it on one screen. The following is a description of the navigation tools. Click the dropdown arrow next to each icon to see options.

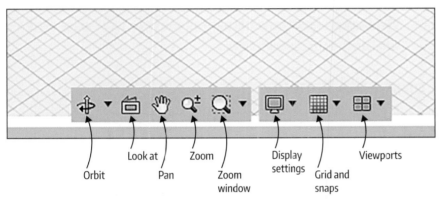

FIGURE 1-14: The Navigation bar

▶ **Orbit:** This whirls you, the viewer, around the model so you can see it from all angles and heights. Your position relative to the model moves, not the model itself. You can orbit constrained or free. Constrained orbits around the x-y plane or the z-axis. Free orbits in any direction. A quicker way to free orbit is to hold the Shift key down *first*, and then hold down the mouse's scroll wheel.

▶ **Look At:** Click a spot on the model and then click this icon. The model's position will adjust accordingly.

▶ **Pan:** This moves the model around on the screen. Pan more efficiently by holding down the scroll wheel of the mouse and moving it around.

▶ **Zoom:** This gives you a magnified view of the model (think telephoto lens) to see small details, or reduces your view of it (think wide-angle lens) to see the big picture. Zoom more efficiently by rotating the mouse's scroll wheel up and down.

- ▶ **Zoom Window:** Drag a window around a specific location you want to look closer at. The Fit option fills the screen up with the model. If your model hides off in a corner after you click it, you've got little pieces you drew earlier still hanging around. Find and erase them, and your model will come back. Clicking Fit is a way to locate "lost" pieces.

- ▶ **Display Settings:** Adjust the look of the workspace.

- ▶ **Grid and Snaps:** Make the grid visible or invisible; adjust its settings and the snap increments.

- ▶ **Viewports:** Show the workspace as one large workspace or as smaller, multiple workspaces.

Comments Panel

Type comments here to collaborators (Figure 1-15) and click Post when finished. Click the minus (–) sign to minimize it. Dock it by moving it over a vertical workspace edge until a green line appears, and then release the mouse.

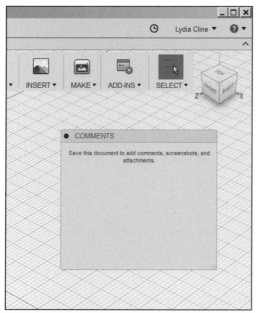

FIGURE 1-15: The Comments bar is for typing notes to collaborators.

THE MARKING MENUS

To perform the same action over and over, use the marking menus. Right-click, hold, and drag the mouse to the left or right to access it (Figure 1-16). This brings up a menu that includes a command for the last operation performed.

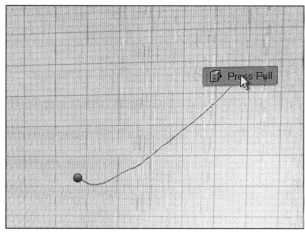

FIGURE 1-16: Right-click, hold, and drag the mouse left or right to access the marking menus.

COLLABORATION

Log on to **a360.autodesk.com** to access your cloud-based workspace (Figure 1-17). Here's where your files are stored, and you can access them from your desktop computer, mobile device, or the web. You can share and discuss your work with others You can also rename the files. Figure 1-17 shows my workspace; the files are on the Data tab. Click the Calendar tab to enter important dates; click the Discussions tab to post notes; or click the Invite button to bring other people into the project.

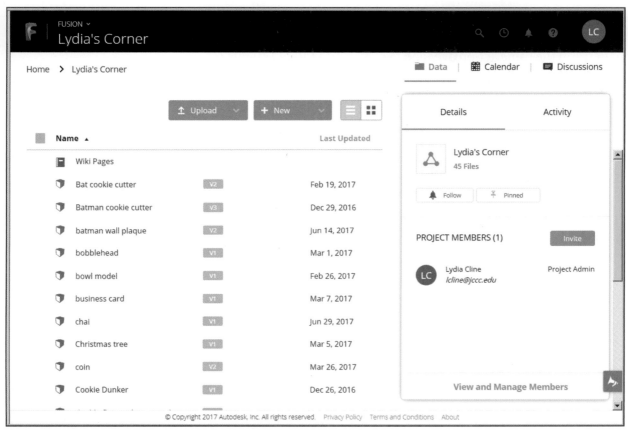

FIGURE 1-17: The online workspace

Saving Files On- and Offline

To save a file, click File/Save or File/Save As. This will save your file to Autodesk's servers, and you can access them by clicking the Data panel icon or going directly to your A360 account. You can also save your files offline by clicking File/Export and checking Save To My Computer (Figure 1-18). You'll be able to work on that file offline and send it to others. Note the New Design From File choice. That's how you open a Fusion or any other supported file that's stored on your computer. You can also export any drawing you make as a DXF file and print it. Many free DXF viewers are available, and they'll let you print the drawing to scale.

FIGURE 1-18: Save files offline with the Export function.

Screenshots

There is no option to send the file to a traditional printer. If you want paper copies of your model, click File/Capture Image (Figure 1-19) to make a screenshot. Alternatively, make a screenshot on a PC by typing **snip** in the Start menu search field to bring up the snipping tool, and then dragging the cursor around the area you want to screenshot. On a Mac, press Command+Shift+4, which temporarily replaces the cursor with a selection window, and then click and drag the tool to make your selection. The screenshot will appear on the desktop.

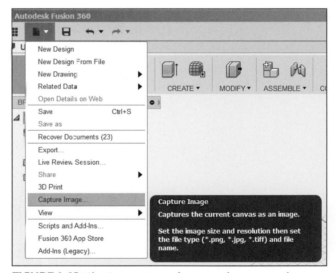

FIGURE 1-19: The Capture Image feature makes a screenshot.

GRAPHICS DIAGNOSTIC TOOL

If Fusion is running slowly on your device, check out the Graphics Diagnostic tool. Click Help/Graphics Diagnostic and a window will appear, showing the current settings (Figure 1-20). You might be able to adjust them for better performance. There's also a Copy To Clipboard button for pasting the graphics information into a word processing program.

FIGURE 1-20: The Graphics Diagnostic tool

And now that you've been introduced to Fusion 360, join me in Chapter 2, where you'll learn how to sketch!

Additional Resources

To see a list of menus, tools, and what they do:

https://help.autodesk.com/view/fusion360/ENU/?guid=GUID-4A759F1B-46A9-49B3-A602-C7A9F20EF947

To see the Fusion Shortcuts and Hotkey Guide:

www.autodesk.com/shortcuts/fusion-360

2

SKETCHING

In this chapter you'll learn how to sketch, a fundamental modeling skill. Sketches are what make parametric models parametric. We'll look at the work plane and its settings, edit and import sketches, understand the difference between driving and driven dimensions, and discuss constraints.

THE WORK PLANE

The work plane is the gridded surface on which you sketch (Figure 2-1). There's a horizontal plane and two vertical planes. You can also make a work plane at any angle of your choice. The first step to sketching is to pick the work plane that you want to draw on. When you click a sketch or solid (the latter, a 3D form, is also called a body), the origin planes and axes will appear. The work plane is a Cartesian system and the circle in the center is the origin (point 0 0,0).

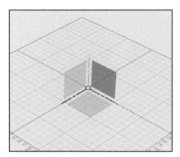

FIGURE 2-1: The work plane is a gridded surface on which you sketch. The origin planes appear when you start to sketch.

GRID SETTINGS

By default, the grid is adaptive, meaning that when you zoom in and out, the grid spacing changes accordingly. But grid settings can be changed. For example, to give the grid fixed numbers, click the Grids And Snaps icon on the Navigation panel at the bottom of the screen (Figure 2-2). Click Set Increments → Fixed, and set the major grid space and number of subdivisions wanted.

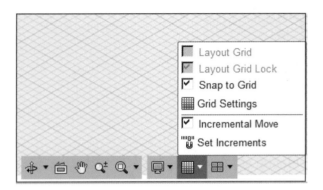

FIGURE 2-2: The Grids And Snaps options

Snapping restricts movement to specific intervals. Sketches and bodies snap to the grid lines, enabling you to model accurately. You can set increments of your choice for linear and rotational moves. You can also turn the incremental move feature off completely to glide across the workspace. This is useful when a tool keeps snapping past the point you want. Snapping is important when you are sketching a shape because if it doesn't snap to a grid or drawn line, you may not have drawn a closed shape (more about that later).

UNITS

Millimeters are the most common unit for 3D printing and are Fusion's default. But you can change the units in a couple of ways. One way is in the Preferences menu; click the dropdown arrow next to your name and then click Default Units → Design (Figure 2-3). The change will take place on a new design, not the current one, so click File → New Design to work in those units. Alternatively, click the Units field in the Browser (it is under Document Settings), and then click the graphic on the right (Figure 2-4). Choose the new units from the dialog box that appears. The units change will take place right away on the open file.

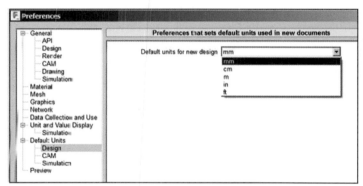

FIGURE 2-3: Change units in the Preferences menu (left two graphics) or in the Browser (right graphic).

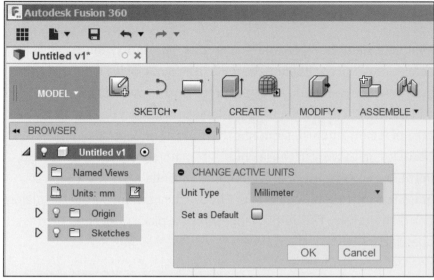

FIGURE 2-4: Change the units mid-design using the Units box in the Browser.

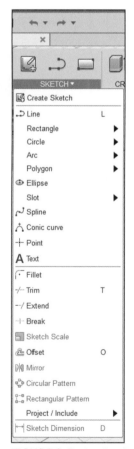

FIGURE 2-5: The Sketch menu

THE SKETCH MENU

Click the Sketch menu's dropdown arrow to see all the choices (Figure 2-5). When you hover over a tool, pop-ups appear that briefly describe how to use it. Some tools have fly-out arrows that access submenus.

Some of this menu's tools, like *Rectangle, Circle, Ellipse*, and *Polygon*, are premade shapes. Others, like *Spline* and *Arc*, let you draw your own shapes. *Fillet, Trim,* and *Extend* edit sketches; *Offset* copies and places a sketch a specified distance from the original; *Mirror* reverses the sketch's position; *Circular Pattern* and *Rectangular Pattern* copy and arrange the sketch; and *Project* puts a sketch onto another surface.

SKETCH SOME SHAPES

After you click a Sketch tool, you make most sketches with three clicks. The first selects the sketch plane, the second selects its starting point, and the third selects its endpoint or size. Click the Line tool. Three picture planes will appear (Figure 2-6); click the plane you want to sketch on. I clicked the horizontal plane. It will change position on the screen to face you. Click a start point and then draw a shape (Figure 2-7). Press Escape or Enter, or right-click the line and choose Cancel to cancel the tool without exiting the sketch. By "without exiting the sketch," I mean all sketch curves will be drawn on the same sketch (more about this in a bit). Click the Line tool again to resume drawing.

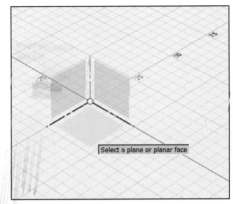

FIGURE 2-6: Click a plane to sketch on.

First point

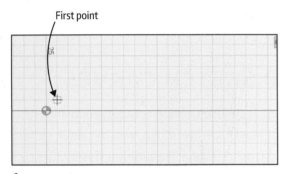

1 Click first point

Second point

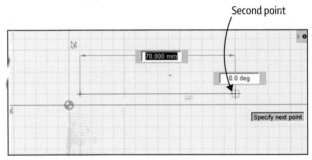

2 Click second point

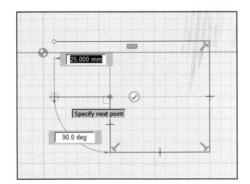

3 Click rest of the points

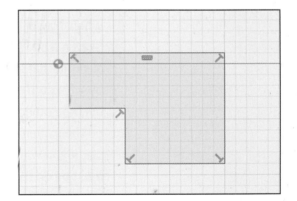

A closed sketch

FIGURE 2-7: Draw a shape with the Line tool. The fourth graphic shows a closed-loop sketch.

When all the endpoints are connected a face, or darkened surface, forms between its lines. This is called a closed-loop, or closed, sketch. An open sketch has nonconnecting endpoints, and therefore, no face. Most operations must be performed on a closed sketch.

When you're finished with the sketch, click the Stop Sketch icon in the upper-right screen, or right-click and choose OK (Figure 2-8).

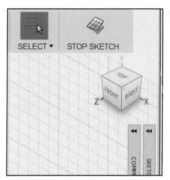 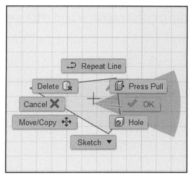

FIGURE 2-8: To finish a sketch, click Stop Sketch or right-click and choose OK.

SKETCHES AND SKETCH CURVES

The terms *sketch* and *sketch curve* are used a lot. A complete drawing is a "sketch" and segments of that sketch are "sketch curves." A sketch curve can actually be straight, curved, or any shape; it's simply a line segment. Some operations work only on specific sketch curves, not the whole sketch.

Sketch curves make up a sketch. They're all in the same sketch when drawn during one operation—that is, before you click the Stop Sketch icon. You can add new sketch curves to an existing sketch by clicking that sketch instead of clicking the grid. It's not visually obvious if sketch curves are on the same or different sketches, but it will become obvious when you work with them. All sketch curves must be on the same sketch to be edited and operated on as a whole. For example, if you want to trim overlapping sketch curves, those sketch curves must be on the same sketch; if they aren't, you can't trim them. Sometimes you'll want sketch curves on the same sketch, sometimes you won't; it depends on what you want to do.

INFERENCES AND CONSTRAINTS

Inference lines and symbols appear while you sketch. These are small geometric shapes that indicate, for example, center points or perpendicularity, as well as dashed lines that indicate that a line you're currently drawing matches the length of an existing line. In the sketch we just drew, you can see perpendicularity, midpoint, endpoint, and dashed line inferences (Figure 2-9).

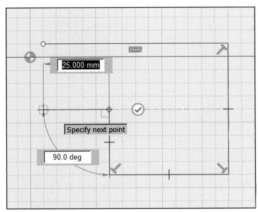

FIGURE 2-9: While drawing this sketch, various inferences appeared.

FIGURE 2-10: A constraints palette appears when you start sketching.

The sketch in Figure 2-9 also shows symbols in the corners called geometric constraints. These are limitations you impose on the sketch's shape. Geometric constraints enforce relationships between parts of the shape. For instance, you can constrain a drawing so that two lines are always parallel or perpendicular to each other. A geometric constraints palette appears when you start sketching (Figure 2-10). Click a constraints icon and then click two sketch items to apply it. The lines will adjust accordingly. To exit from applying a constraint, press the Escape key or right-click and choose Cancel.

Sketch dimensions must be visible for constraints to work properly; if they're not, the constraints will do things you probably don't want, like stretch one sketch curve to another. The dimension sketch mode is discussed later in this chapter.

A drawing can be fully constrained (also called fully defined) with constraints and dimensions. It can also be partly constrained or completely unconstrained. To prevent constraints from forming, press and hold the Ctrl key while drawing each sketch line. To delete a constraint, click the constraint icon and delete. A blue circle will appear to show that it is selected (Figure 2-11). Press the Delete key. The constraint will no longer apply to that sketch line. If a constrained sketch is inside a body, turn the body off through the Browser first (click the light bulb in front of it to make it gray instead of yellow), and then delete the constraint.

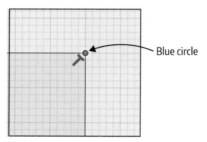
Blue circle

FIGURE 2-11: To remove a constraint, select and delete it.

A fully constrained sketch means you can click any point on it to move it and it will move as a whole without changing its shape. None of its lines can move independently. When a sketch isn't fully constrained, part of the sketch will move and change. The advantage of a fully constrained sketch is that it will always maintain its shape (design intent) and produce predictable results when you perform other operations on it such as mirroring its features.

> **TIP** If trying to constrain two sketches doesn't work—for example, one won't move to the other in the desired order, or the sketch changes sizes— try applying the Fixed constraint to one and then applying the desired constraint to the other.

SELECT AND DELETE SKETCHES

There are multiple ways to select an item. You can:

- ▸ Drag a selection window. This is from upper left to lower right and selects everything that is entirely inside the window.

- Drag a crossing window. This is from lower right to upper left and selects everything the window touches.

- Click its Browser entry. This always selects the entire item.

- Click its Timeline option. This always selects the entire item.

- Click options under the Select menu that's at the top of the screen.

Selected items appear dark blue. Selecting isn't always a simple task. Depending on how you select something, you will get different context menus. Or you may not have selected enough of it to perform an operation; this is more an issue with bodies. You may not have selected the body's under or back sides.

Click the dropdown arrow in front of the word Sketches to see all sketches in the file. A gray bulb in front of the sketch name means the sketch is invisible. Click the bulb to turn it yellow, which makes the sketch visible. In Figure 2-12, I selected the sketch by clicking its Browser panel entry. To select multiple items, press and hold the Shift key and click them.

Simply clicking the face of a sketch won't select it; you have to select all its edges, too. An alternative to the selection window is to select the whole sketch through the Browser panel. In fact, sometimes this is the only way to delete some items, such as a sketch point or a sketch curve if just selecting, right-clicking, and choosing Delete doesn't work.

You can generally drag a selection window around a sketch and then press the Delete key to remove it. You can also right-click the sketch and select Delete. If the sketch still won't delete, select it, right-click, and choose Edit Sketch from the context menu. Then select it by dragging a window around it and press Delete.

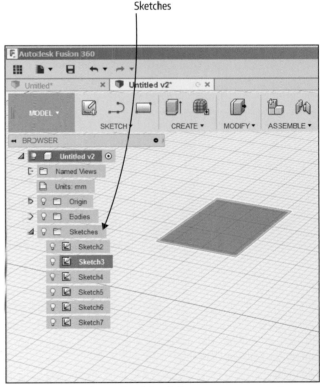

Sketches

FIGURE 2-12: Select a sketch by clicking its Browser panel entry.

CREATE A SELECTION SET

You can select lines and save that selection to revisit for future editing (Figure 2-13). Select the lines (press and hold the Shift key to make multiple selections), right-click one, and choose Create Selection Set. This selection will show up in the Browser panel, where you can rename it. Hover over the selection's Browser panel entry to access two icons; one lets you select all those lines again, and the other lets you update any changes to the selection set.

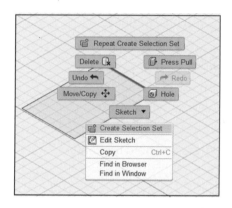 

FIGURE 2-13: A selection set may make future editing easier.

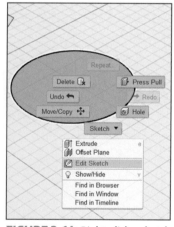

FIGURE 2-14: Right-click a sketch and choose Edit Sketch to enter Edit mode.

EDIT A SKETCH

Once you click the Stop Sketch icon, you leave the sketch. To change it later, you need to enter Edit mode. Right-click a line or face of the sketch to bring up a context menu, and then click Edit Sketch (Figure 2-14). After editing, click Stop Sketch. Note that if you can't edit a sketch, it might be fixed (locked), as discussed later in this chapter.

TIP If the grid becomes oriented differently from the sketches, enter Edit mode on a sketch, and then exit it. The grid will orient to that sketch.

MOVE, ROTATE, AND COPY A SKETCH

Moving versus copying involves different steps, even though there is one Move/Copy option on the context menu.

Moving a Sketch

Select a sketch, right-click, and choose Move/Copy. Typically, a sketch's face and all edges must all be selected; if they're not, a dialog box will prompt you to select them (Figure 2-15). Once selected, a widget will appear (Figure 2-16). Sometimes, however, you just need to click the edge. Let what appears guide you; if the widget doesn't appear when a sketch is totally selected and the dialog box won't let you make a selection, then just select the sketch's edges. This is often the case with circle sketches.

> **TIP** Dialog boxes that appear during an operation have an *i* in the lower-left corner that you can click to access a tip box.

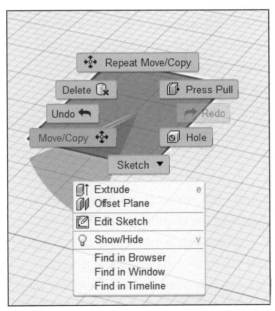
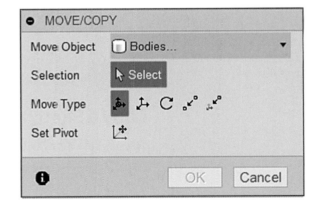

FIGURE 2-15: If the sketch isn't entirely selected before choosing Move/Copy, a box will appear with a Select button. Click the button if it isn't highlighted already, and then select the whole sketch.

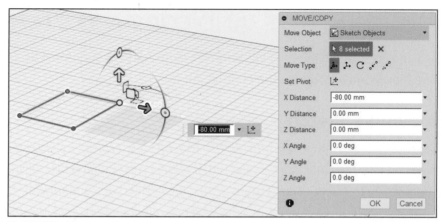

FIGURE 2-16: Move the sketch with the widget.

The widget can move, rotate, and copy. It has nine manipulators: three arrows, three planes, and three buttons. Drag the arrows to move a selected item up and down, back and forth, or left and right along the arrows' respective axes. Drag the planes to move an item freely up and down, back and forth, or left and right along the planes' respective axes. Drag the buttons to rotate an item up and down, back and forth, or left and right along the buttons' respective axes. You can snap the item along the grid as you move it, type a specific amount in the text field, or just "eyeball" a new position. Note the row of icons where you can choose specific move types: free move, translate, rotate, point to point, and point to position (Figure 2-17). If you can't move a sketch, it might have a constraint somewhere on it, usually in the middle. Select and delete it.

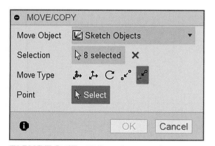

FIGURE 2-17: Click an icon to choose a specific type of move.

Copy a Sketch

To copy a sketch, enter Edit mode. Next, select it and press Ctrl+C and then Ctrl+V. A copy will be made over the original item and the widget for it will appear. Move the copy where you want. Repeat this process for each new copy. Alternatively (while also in Edit mode), you can select the item, right-click, choose Copy from the context menu, right-click somewhere on the grid, and choose Paste (Figure 2-18). The widget will appear; move it where you want the copy to be.

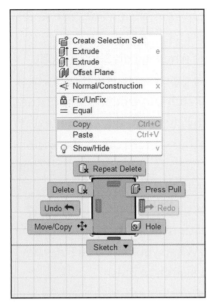 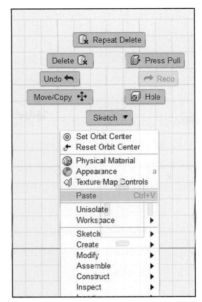

FIGURE 2-18: While in Edit mode, right-click, choose Copy, and then right-click the grid and choose Paste.

> **TIP** Use the Rectangular Pattern tool to make quick copies. In fact, it's best to use the pattern and mirroring tools when possible instead of drawing more complex sketches.

DIMENSIONS AND SKETCH EQUATIONS

When sketching, you can eyeball proportions if you're just trying to figure out a design or count grid squares and click sketches onto the grid lines. This offers precision but not the tedium of typing dimensions all the time. Starting a sketch at the origin makes counting grid squares easier. Changing your grid spacing to accommodate the grid lines for what you are trying to do is useful; if you know that your snap points will be a specific interval, you can set that.

Type dimensions when you need to be specific. These control the size and distance of objects. Text fields appear while you sketch, and the Tab key toggles between them. After you type a dimension, a dimension line and note will appear on the sketch, and this also adds a constraint. When you click Stop Sketch, the dimension lines and notes will become invisible.

Dimension lines and notes are visible when in Edit Sketch mode, and they need to be visible if you want to type sketch equations. These are formulas that make the size of one sketch curve dependent on the size of another. Create this relationship by clicking a dimension text field to activate it, and then clicking another sketch curve's dimension note (don't click the sketch curve itself; click its dimension note). That dimension note will then appear in the activated text field.

Figure 2-19's top graphic shows an activated dimension text field. You can enter dimensions or sketch equations in it. The bottom graphic shows an example of a sketch equation. When you press the Enter key, that line will adjust to 27 mm. Type Stop Sketch to finish.

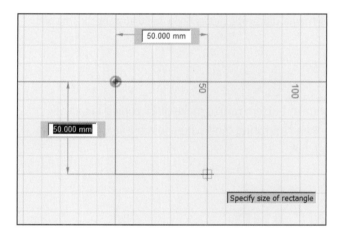

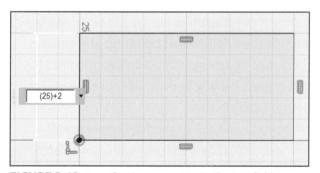

FIGURE 2-19: Type the size or equation in the text field; press Tab to toggle between text fields.

To change dimensions later, enter Edit Sketch mode. Then select a sketch line, right-click that selected line, and choose Sketch Dimension (Figure 2-20) from the context menu. A dimension line will appear; drag it off the sketch. The text field shows the line's current length. Type a new length, if you want. The sketch will resize to that length, and if it has horizontal/vertical constraints, the shape will remain the same (Figure 2-21). You can also add a new unit to the dimension by typing the name of that unit after the number—such as adding *in* for inch to a file that is currently in millimeters—and Fusion will convert the new unit to the file's current units.

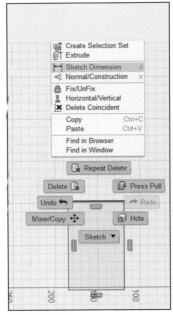

FIGURE 2-20: Choose Sketch Dimension from the context menu.

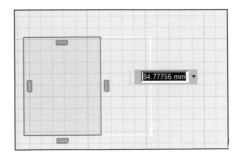

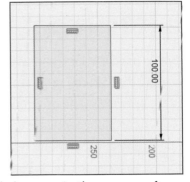

FIGURE 2-21: Because of the constraints, the sketch's shape remains the same even after a dimension was changed.

DRIVING VS. DRIVEN DIMENSIONS

The dimension line drawn in Figure 2-21 is a driving dimension—that is, it drives the size of the model. Adding a lot of dimension lines to a sketch will overly constrain it, making some of those dimension lines driven, or reference lines. A driven dimension tells you how long the line is, but it can't be edited. You have to edit the line itself to change it. Note that if the sketch is unconstrained, changing the dimensions will change the shape (Figure 2-22). A shape changes when a driving dimension tells it to change.

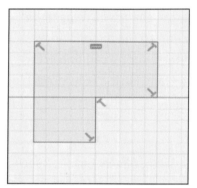 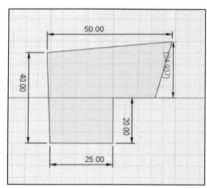

FIGURE 2-22: Changing the dimensions will change the sketch's shape if the sketch is unconstrained.

SKETCH TOOLS

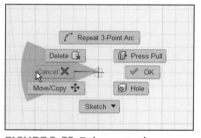

FIGURE 2-23: To leave a tool, press Escape or right-click and choose Cancel.

We've used the Line and 2-Point Rectangle tools so far. Let's see how some other Sketch tools work. Remember that to exit any of them, just press Escape or right-click and choose Cancel (Figure 2-23). You can also reactivate the tool by right-clicking; the top option in the context menu repeats the last tool used. It's good practice to incorporate the origin into a sketch, such as a corner or midpoint. Some tools, such as Sketch Scale, work only when you can click the origin point. However, know that sketching at the origin point often leaves a constraint there, which might interfere with future operations. Delete the constraint by dragging a selection window around it and pressing Delete or by deleting its entry in the Browser panel.

- ▶ **3-Point Rectangle:** This tool in the Rectangle submenu lets you draw rectangles at an angle (Figure 2-24).

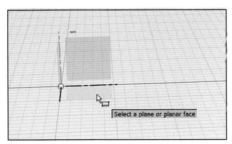

1 Click the plane

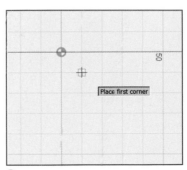

2 Click first point

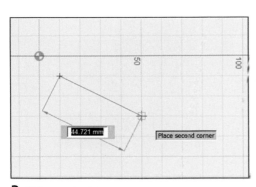

3 Click second point

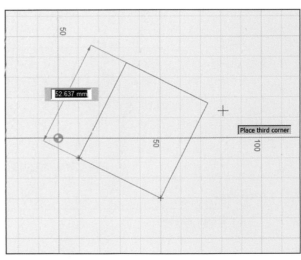

4 Click third point

FIGURE 2-24: The 3-Point Rectangle can draw angled rectangles.

- ▶ **Center Rectangle:** This tool in the Rectangle submenu lets you draw a rectangle by choosing a center and one corner (Figure 2-25).

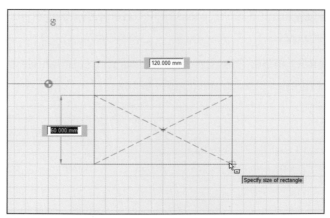

FIGURE 2-25: The Center Rectangle is drawn with a center point and one corner.

▶ **3-Point Arc:** This tool in the Arc submenu lets you draw an arc by choosing two endpoints and then a third for the bend (Figure 2-26).

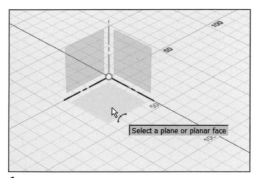

1 Click the plane

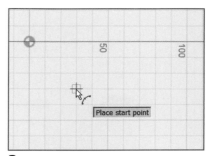

2 Click first point

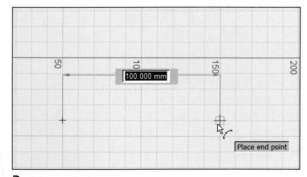

3 Click second point

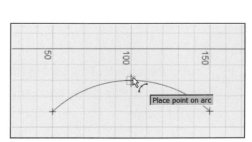

4 Click the bulge

FIGURE 2-26: The 3-Point Arc is drawn with two endpoints and a bend.

► **Tangent Arc:** This tool in the Arc submenu draws an arc that's tangent to an endpoint on another sketch (Figure 2-27) or to a deliberately created point somewhere on that sketch. Click that sketch instead of the grid when clicking the first point to put the arc on the same sketch as the other sketch curves.

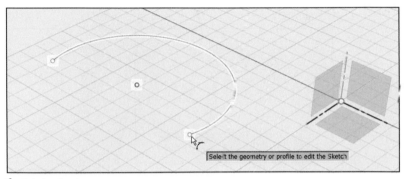

1 Click the sketch

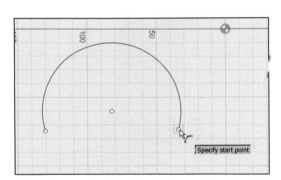

2 Click first point

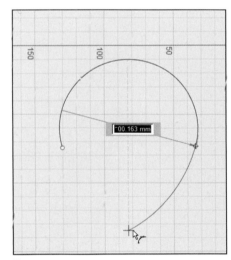

3 Click second point

FIGURE 2-27: The Tangent Arc attaches to another sketch.

▶ **Circumscribed/Inscribed Polygon:** These tools are in the Polygon submenu. One draws a polygon around a circle, the other draws one inside a circle. Click the center and then click the midpoint of an edge or type a distance to the edge (Figure 2-28). The default number of sides is six, but you can click inside the Edge Number field and type a new one (Figure 2-29).

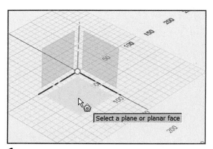

1 Click the plane

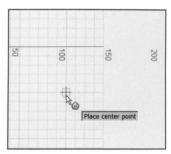

2 Click center point

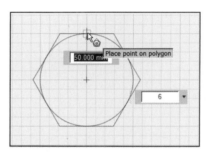

3 Click polygon point

FIGURE 2-28: Circumscribe a polygon by clicking a center and edge.

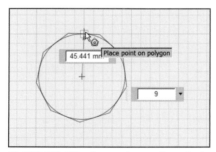

FIGURE 2-29: Change the number of sides by clicking in the Edge Number field.

► **Edge Polygon:** This tool lets you, among other things, match a polygon edge to the size of an existing sketch curve. Click two points on a sketch curve and then click the work plane on the side you want to place the polygon (Figure 2-30). You can click the two points wherever you want, such as on endpoints or on the midpoint and endpoint.

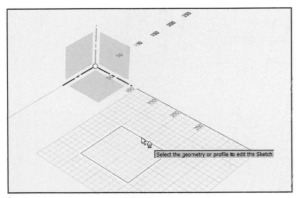

1 Click the sketch

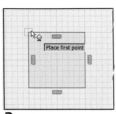

2 Click first point

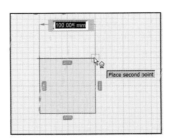

3 Click second point

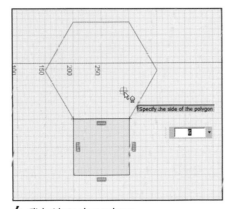

4 Click side to place polygon

FIGURE 2-30: An Edge Polygon lets you match a polygon edge to the size of an existing sketch curve.

Ellipse: This tool lets you draw an oval shape by choosing a center, major axis, and point on the ellipse (Figure 2-31).

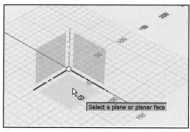

1 Click a plane

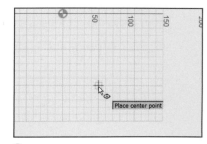

2 Click center point

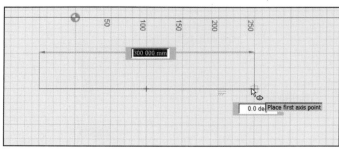

3 Click axis point

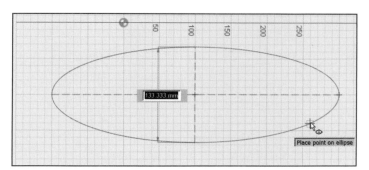

4 Click ellipse point

FIGURE 2-31: The Ellipse tool

> **Slot:** This tool draws a long, narrow slit in which something can be inserted. You can construct it three different ways from options in the Slot submenu; Figure 2-32 shows a slot made in the ellipse we just drew.

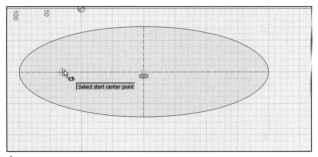

1 Click first center point

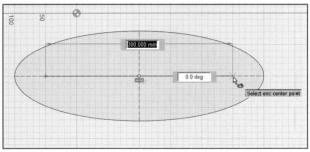

2 Click second center point

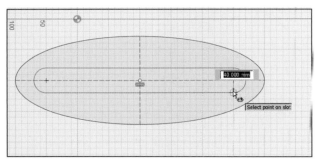

3 Click slot width

FIGURE 2-32: A slot, created with the center-to-center option

▶ **Spline:** This tool creates curves that connect two or more points (Figure 2-33). When you complete a closed loop, tangent points and tangent handles appear. Grab them and drag to change the shape (Figure 2-34).

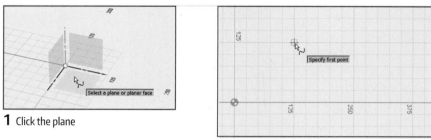

1 Click the plane

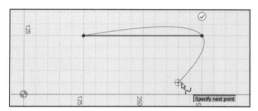

2 Click first point

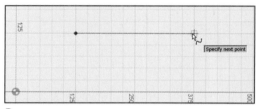

3 Click second point

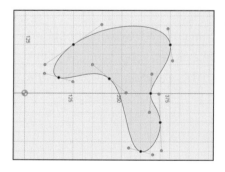

4 Click the spline

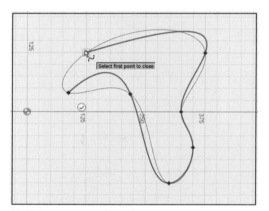

Multiple, connected splines

FIGURE 2-33: The Spline tool creates curves that connect two or more points.

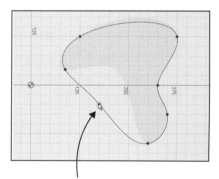

Dragged vertex point

FIGURE 2-34: Grab and drag the tangent points and handles to change the shape.

▶ **Conic Curve:** This tool creates curves similar to splines, but it gives you more control over the exact dimensions (Figure 2-35).

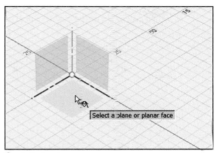

1 Click the plane

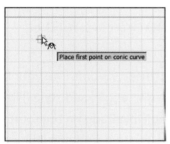

2 Click first point

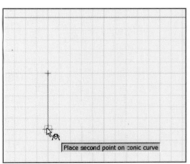

3 Click second point

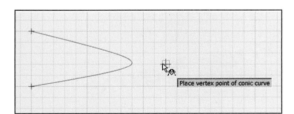

4 Click the vertex point

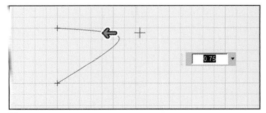

5 Adjust the curve

FIGURE 2-35: Creating a conic curve

▶ **Point:** This tool lets you place points at precise locations (Figure 2-36). You can then snap sketch curves to them, useful for positioning holes. Sketch curves need to be on the same sketch as the points to snap to them.

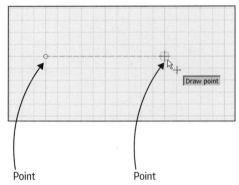

Point   Point

FIGURE 2-36: Click the grid to create points.

▶ **Text:** This tool inserts words into your file. Click the Text tool and choose the height, angle, and style (Figure 2-37). You can also choose the angle by dragging the rotator button that appears when you click the text tool onto the grid.

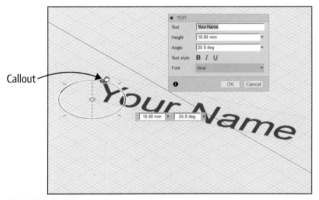

FIGURE 2-37: Click the text tool onto the grid to type words.

▶ **Fillet:** This tool places an arc at an intersection of two lines or arcs. You choose the fillet's radius. Figure 2-38 shows the Fillet tool clicked onto a rectangle's corner.

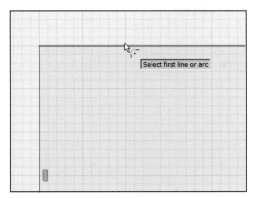

1 Select one line

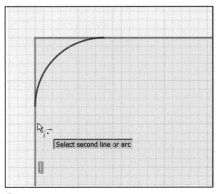

2 Select a second line

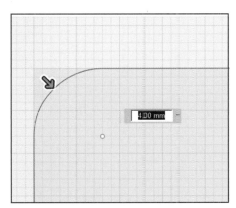

3 Drag the arrow to adjust

FIGURE 2-38: Click two lines. A fillet will appear; drag the arrow or type a dimension to adjust its size.

▶ **Trim:** This tool lets you remove sketch curves between overlapping sketches. In the process it creates a new, separate shape. The sketch curves all have to be on the same sketch. Click the Trim tool onto one of the sketch curves, then hover it over a line you want to remove. It will turn red. Click, and the sketch curve will disappear (Figure 2-39). Clicking trims lines one at a time. Alternatively, press and hold the left mouse button and slide it to trim continuously.

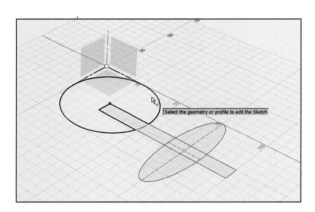

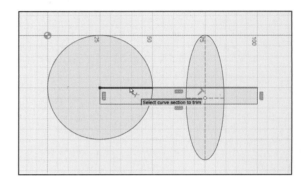

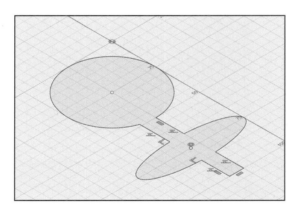

FIGURE 2-39: Trim overlapping sketch curves.

▶ **Extend:** Use this tool to extend sketch lines. If you have a line that's short of where it should be, hover the Extend tool on it and a red line will appear that extends to the nearest sketch line (Figure 2-40). Click to finish.

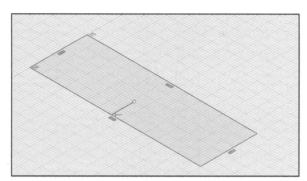 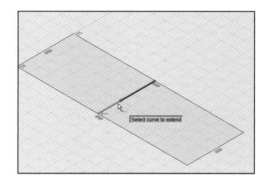

FIGURE 2-40: The Extend tool makes a line longer.

▶ **Break:** This tool lets you cut a continuous line where another line intersects it. Hover over a line to select it. A rec X will appear at the intersection. Click it, and the highlighted line will break into two lines (Figure 2-41).

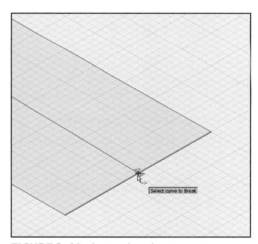

FIGURE 2-41: The Break tool cuts a continuous line into two lines.

- **Sketch Scale:** This tool makes a sketch larger or smaller. You must be in Sketch Edit mode and the sketch must have one point on the origin for this tool to work. So if the sketch isn't on the origin, move it there. Then select the whole sketch, activate the tool, and click the Point button in the dialog box. An arrow will appear. Drag the arrow to scale the sketch or type a dimension in the text field (Figure 2-42).

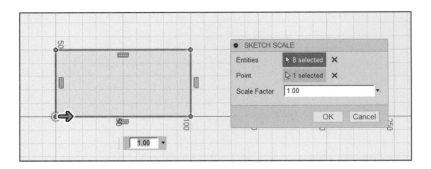

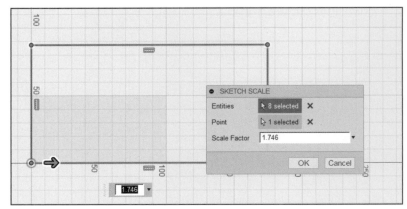

FIGURE 2-42: The Sketch Scale tool makes a sketch larger or smaller.

- **Offset:** This tool makes a copy of the sketch a specified distance from it (Figure 2-43). Click the sketch you want to offset, click the specific sketch curve(s) to offset (press and hold the Shift key to make multiple selections), and then drag the arrow or type a specific offset distance.

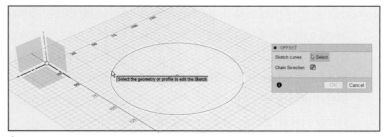

1 Click the sketch

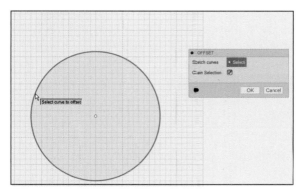

2 Click the circle

Handles

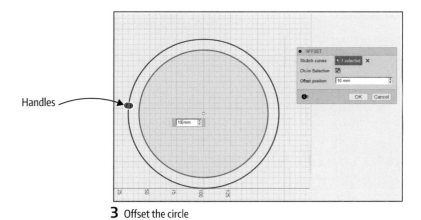

3 Offset the circle

FIGURE 2-43: Offsetting a circle sketch

▶ **Mirror:** This tool makes a reversed copy of a sketch curve. You need a mirror line, and this line must be on the same sketch as the sketch curve being mirrored. Select the sketch curve, select the line to mirror about, and click OK to finish (Figure 2-44). An inability to mirror probably means the mirror line is not on the same sketch as the sketch curve being mirrored.

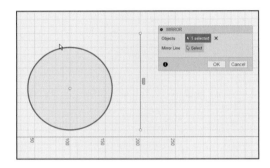
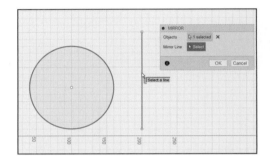

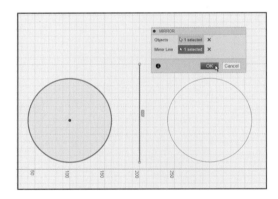

FIGURE 2-44: Mirroring a sketch about a line. The line and sketch must be on the same plane.

▶ **Circular Pattern:** This tool arrays (copies and arranges) a sketch curve or whole sketch around a circle. It can array completely around the circle or just partially. The sketch must be in Edit mode.

Select the sketch, activate the Circular Pattern tool, click the
Center Point button in the dialog box, and then click a center
point (the point around which the sketch is arrayed). Handles and
two copies will appear. Drag the handles forward to add copies;
drag them backward to subtract copies. Or just type the number
of copies wanted in the text field. Click OK to finish (Figure 2-45).
If you later change the dimension of one sketch, all the arrayed
copies will change, too.

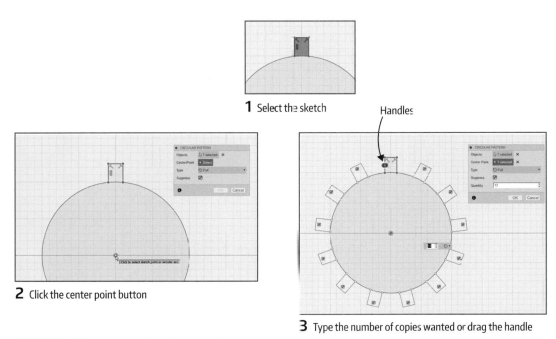

1 Select the sketch

Handles

2 Click the center point button

3 Type the number of copies wanted or drag the handle

FIGURE 2-45: Arraying a sketch in a circular pattern

▶ **Rectangular Pattern:** This tool arrays (copies and arranges) a
  sketch in a row or column (Figure 2-46). The sketch must be in
  Edit Sketch mode. Select the sketch and then activate the Rectangular Pattern tool. Drag the arrows back and forth and up and
  down to create rows and columns. Alternatively, type how many
  rows and columns you want. Note all the options in the dialog
  box. You can specify one direction or symmetric, the distance
  between copies, and the distance type (extant or spacing). Click
  the checkmarks to remove copies.

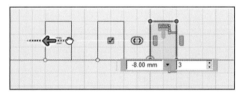

1 Select the sketch

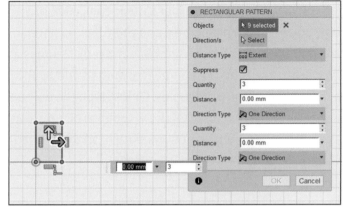

2 Grab an arrow

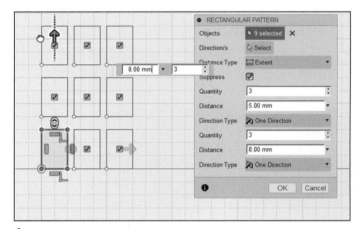

3 Drag the arrow left or right

4 Drag the arrow up and down

FIGURE 2-46: Arraying a sketch in a rectangular pattern

▸ **Project:** This tool creates a new sketch from selected sketch curves or edges of bodies. The following are steps for projecting a sketch to a block:

1. Click Project / Include → Project To Surface (Figure 2-47).

2. Click a horizontal plane because you need to tell the software which direction to project. I clicked the origin plane (Figure 2-48), but you could click a horizontal surface on the block, too. A dialog box will appear and the sketch will orient as a top-down view. Click the ViewCube's house icon to return to a 3D one.

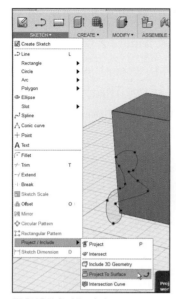

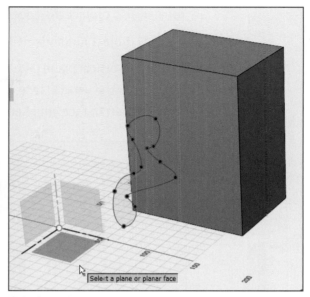

FIGURE 2-47: Click Project / Include → Project To Surface.

FIGURE 2-48: Click a horizontal plane.

3. In the dialog box, the Faces button is highlighted. Click the block's front face. Then click the Curves button to highlight it and select the sketch curve. It will project forward to the block (Figure 2-49). Click OK to finish.

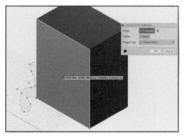

1 Select the face

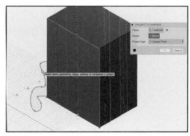

2 Select the sketch

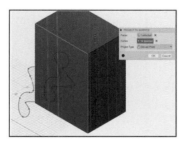

The projected sketch

FIGURE 2-49: With the Curves button selected, click the sketch curve. It will project to the block.

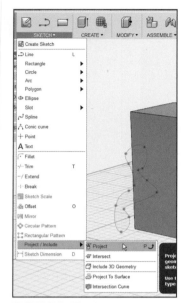

FIGURE 2-50: Click Project / Include → Project.

You can also project the edges of the block forward to the sketch.

1. Click Project / Include → Project (Figure 2-50).

2. Click a vertical plane (Figure 2-51), either the vertical origin plane or the vertical face of the block. The sketch and block will orient to face you. Keep this orientation.

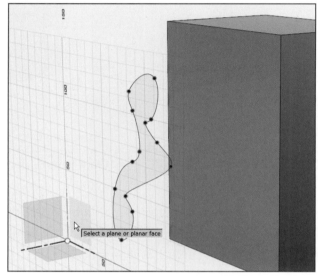

FIGURE 2-51: Click a vertical plane.

3. Click one of the block's edges. This will select all the edges and they'll project forward to the sketch (Figure 2-52). Click OK in the dialog box to finish.

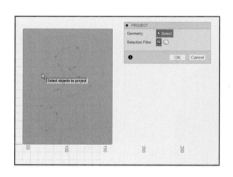

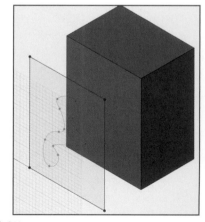

FIGURE 2-52: Click the block's edges and then click OK.

SKETCH ON AN ANGLED PLANE

Earlier we sketched on the origin planes, all of which are perpendicular to each other. However, you can sketch at angles, too. Click Construct → Plane At Angle (Figure 2-53). This creates a construction plane through an edge, axis, or line, at any angle you choose. Click any edge—it could be the edges between the origin planes or the edges on a body. A handle will appear, which you can drag to the desired angle, or just type the angle in the text field of the dialog box that appears. Click OK and you'll be left with the angled sketch plane to draw on.

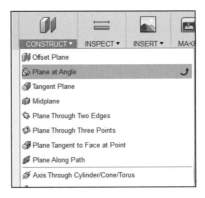
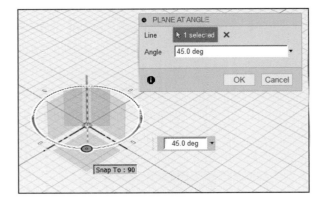

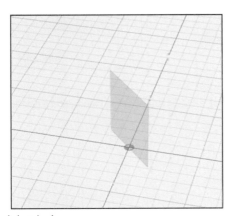

FIGURE 2-53: Creating an angled sketch plane

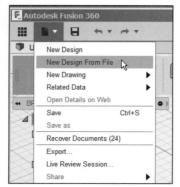

FIGURE 2-54: The Data panel icon and New Design From File

INSERT SKETCHES INTO FUSION

You don't have to draw everything from scratch. You can insert many file formats into Fusion ready to edit. There are three ways to do this: insert into the workspace, upload to the Data panel, or upload via the New Design From File option. Most files must be brought in through the latter two (Figure 2-54) to be translated into the Fusion format.

Insert into the Workspace

DXF (Autodesk Drawing Interchange Format) and SVG (Scalable Vector Graphics) files insert directly into the workspace, ready to edit. Find these functions under the Insert menu.

Let's insert an SVG file. Click Insert → Insert SVG. The origin planes and a navigation folder will appear (Figure 2-55). Click the plane you want to draw on; then click the file folder graphic, find the SVG, and press Enter. The file will appear with a widget and dialog box with which you can move and scale it (Figure 2-56).

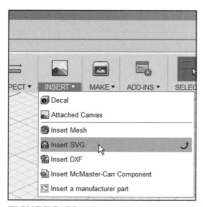

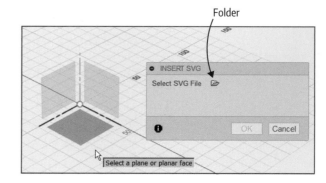

FIGURE 2-55: When you click Insert, an SVG file, the origin planes, and a navigation folder appear.

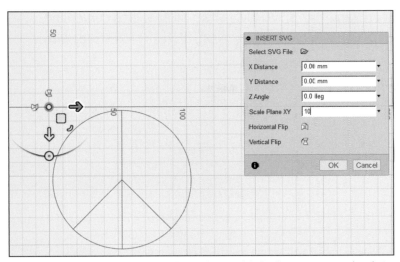

FIGURE 2-56: The file appears with a widget and dialog box to move and scale it.

Fixed/Unfixed Sketches

Inserted sketches usually come in fixed, meaning they're fully constrained, the equivalent of locked and uneditable. You can tell a fixed sketch by its green color. It will also have an orange pushpin next to its entry in the browser panel. Select the sketch, right-click, and choose Fix/Unfix, and the sketch will turn blue (Figure 2-57). In fact, this is something to check if you can't edit any sketch.

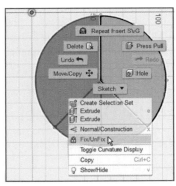
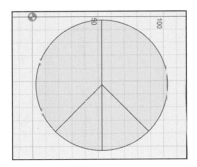

FIGURE 2-57: A green (locked) sketch is fixed; unfix it to turn it blue.

After bringing in this SVG file, I offset all its sketch curves and trimmed overlapping ones (Figure 2-58). It's ready to extrude into a body now.

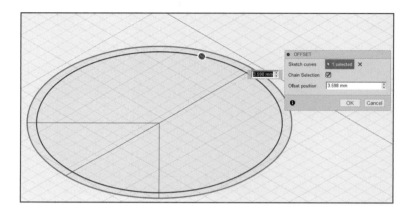

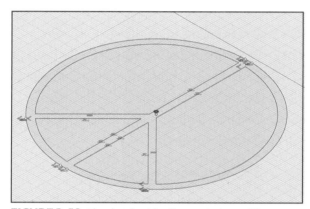

FIGURE 2-58: The sketch with offset and trimmed lines

Upload to the Data Panel

The Data panel brings in these files:

Autodesk Alias (`.wire`)

AutoCAD DWG Files (`.dwg`)

Autodesk Fusion 360 Archive Files (`.f3d`)

Autodesk Fusion 360 Toolpath Archive Files (`.cam360`)

Autodesk Inventor Files (`.ipt`, `.iam`)

CATIA V5 Files (`.CATProduct`, `.CATPart`)

DXF Files (`.dxf`)

FBX (`.fbx`)

IGES (`.ige`, `.iges`, `.igs`)

NX (`.prt`)

OBJ (`.obj`)

Parasolid Binary Files (`.x_b`)

Parasolid Text Files (`.x_t`)

Pro/ENGINEER and Creo Parametric Files (`.asm`, `.prt`)

Pro/ENGINEER Granite Files (`.g`)

Pro/ENGINEER Neutral Files (`.neu`)

Rhino Files (`.3dm`)

SAT/SMT Files (`.sab`, `.sat`, `.smb`, `.smt`)

SolidWorks Files (`.prt`, `.asm`, `.sldprt`, `.sldasn`). Note that Solid-Works drawings (.slddrw) cannot be opened in Fusion 360.

STEP Files (`.ste`, `.step`, `.stp`)

STL Files (`.stl`)

SketchUp Files (`.skp`)

New Design From File

The New Design From File menu brings in these files:

Autodesk Fusion 360 Archive Files (`.f3d`)

IGES (`.ige`, `.iges`, `.igs`)

SAT/SMT Files (`.sat`, `.smt`)

STEP Files (`.step`, `.stp`)

CONVERT A JPEG TO AN SVG

FIGURE 2-59: Import many file formats using the Data panel icon and New Design From File menu.

Fusion can import a JPEG file through the Insert → Attached Canvas menu (Figure 2-59). This puts it on a sketch plane, where you can trace it. However, a more efficient method is to import an SVG file, which comes in as Fusion geometry. This works well for detailed sketches and text. There are online converters, such as the one at **online-convert.com**, that convert JPEGs to SVG files (Figure 2-60). Be aware that some files convert better than others—crisp, clear, black-and-white ones convert the best—and their editability inside Fusion varies. Inkscape, a free digital imaging program you can download at **inkscape.org**, also converts files with its Trace Bitmap tool. Some files convert better with Inkscape than an online converter; others convert better online. If a converted image imports poorly, you can import a JPEG, PNG, or TIFF file by choosing Insert → Attached Canvas and trace on top of it to create sketch curves.

Now that you know how to use all these sketch tools, let's start making models.

FIGURE 2-60: At online-convert.com you can change JPEG files to SVGs.

MAKING MODELS: THE CREATE MENU

In this chapter we'll use the Create menu in the Model, Sculpt, and Patch workspaces to make solid, T-spline, and surface models. Model mode makes solid, precise forms with defined edges and sizes. Sculpt mode makes solid, organic, flowing forms. Patch mode makes hollow, precise forms with defined edges and sizes. We'll bounce models between all three workspaces to take advantage of their unique capabilities.

DIRECT VS. PARAMETRIC MODELING

Parametric modeling works best for projects that need exact dimensions for manufacturing. It uses feature-based solid and surface design tools. All actions are captured in a timeline that allows you to change any operation, such as a sketch's size, and have everything upstream change around it. You can set parameters and relationships between parameters; for example, make one body consistently half the size of another. This lets you automate repetitive changes and make small, customized changes to a basic design.

Direct modeling works best for conceptual thinking. It's easier to design in because there is no associativity between sketches and bodies; that is, you won't get error messages telling you that you can't move a body because a sketch is referencing it. There is no timeline, so you can't change something and have everything around it change, too.

To model directly, right-click the file's title in the Browser and click Do Not Capture Design History (Figure 3-1). A message will appear saying that all design history will be removed; click Continue. Any timeline you have will be removed, and there won't be a timeline from that point forward. You can also model specific operations on the timeline directly; just right-click the operation's timeline icon and choose Convert To DM Feature. That way, you won't lose all the other timeline icons.

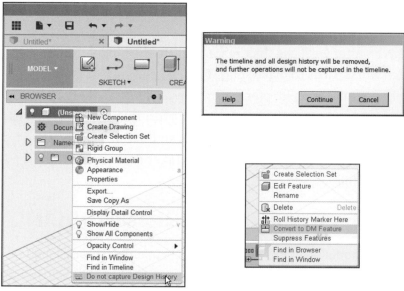

FIGURE 3-1: Turn off design history to model in direct modeling mode. You can turn off the history entirely or turn it off on just one item.

Some tools exist only in direct modeling, and others exist only in parametric. The Model and Patch workspaces offer direct and parametric modeling. Sculpt mode offers only direct.

MODELING WORKFLOW

You can start a project in Model and finish it in Patch, or start it in Sculpt and bring it into Model. This workflow lets you take advantage of each workspace's capabilities. You can bring a project from Sculpt into the Patch or Model workspaces and back again in direct modeling mode only, not parametric. You can't modify a solid model in Sculpt.

THE MODEL WORKSPACE

Click the Create menu (Figure 3-2). Here you'll find ready-made bodies: box, cylinder, sphere, torus, coil, and pipe. To see how one works, click Box and then click an origin plane. Sketch the box with three clicks and pull it up to the desired height (Figure 3-3). Try out the rest of the forms (Figure 3-4). The pipe needs a predetermined line to pull it along; you can sketch a line or spline, or use the edge on a body.

FIGURE 3-2: The Create menu

1

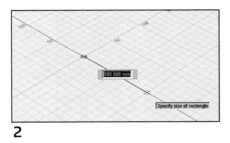

2

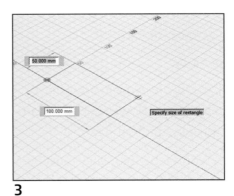

3

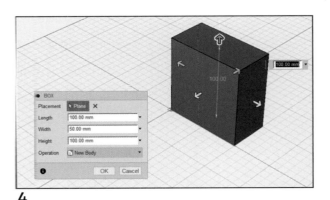

4

FIGURE 3-3: Making a box

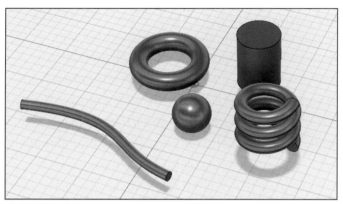

FIGURE 3-4: Ready-made bodies

BODIES AND COMPONENTS

Fusion 360 uses a system of bodies and components. Design decisions should be made based on this system, so let's discuss this topic before going further.

Bodies

A body is a single 3D form. The box we just made is a body. Most tools in the Create menu create new bodies when you choose New Body from their dialog boxes. When you split a body, both parts are bodies. Add to and subtract bodies from each other to make your design. For example, you can model a business card by making a body for the base and a body for a logo, and then joining them together.

The four different types of bodies are as follows:

- ▶ Solid, made in Fusion's Model workspace
- ▶ Surface, made in Fusion's Patch workspace
- ▶ T-spline, made in Fusion's Sculpt mode
- ▶ Mesh, which is an imported STL or OBJ file. A mesh body must be converted to solid in order to be edited. However, Fusion can only convert meshes that have no holes or other flaws.

Different body types can't interact with each other. They must be converted to the same type before you can join, cut, or intersect them. Solid and Surface are Fusion's core body types, so it's best to eventually convert your model to one of those formats. This enables you to use the full Fusion toolset for assembly, manufacturing, and simulation.

All body conversions must be done in direct modeling mode, so if you're in parametric mode, right-click the Browser's title field and choose Do Not Capture Design History. Then select the item to be converted by dragging a selection window around it, right-clicking, and choosing either Mesh To BRep or BRep To Mesh (Figure 3-5). If those options don't appear, try selecting it a different way—for example, by clicking its Browser entry.

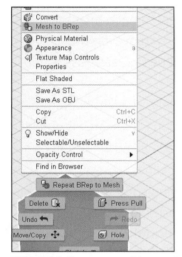
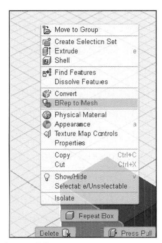

FIGURE 3-5: Convert solid to mesh and mesh to solid.

You can convert:

- Solid/Surface to T-spline
- T-spline to Solid/Surface
- Mesh to T-spline
- Mesh to Solid/Surface (if the mesh has no flaws)

Components

A component is a container that holds a body. More typically, a component contains multiple bodies, sketches, and any other construction objects that make up a part. Components are required for assemblies, CNC instructions, and scaled drawings, and to create a bill of materials. By default, the Browser represents a single component and the top Browser node (the text field with the file's title) is the design's "root." Through the Browser, you can drag one component into another component, which creates a subassembly (a component within a component).

To create a component, right-click the Browser's title field and choose New Component, or click the Assemble menu and choose New Component (Figure 3-6). Those two methods will store the component's parametric history completely inside the component, even when you do a Save As. Other component creation methods, such as right-clicking bodies and choosing Create Component From Body, lose all timeline information. After you create a new component, click the radio button next to it to activate it. When activated, everything— sketches, bodies, construction geometry, joint origins, etc.—is created in that component.

Create components before you start modeling because doing so preserves a complete timeline of your operations. If a sketch is outside the component, you've lost the parametric ability. And when you convert bodies to components, you lose their timelines.

Fusion 360's timeline shows only the component, not everything in it, which makes a quickly growing timeline easier to work with. Make sure the component is active when adding more items to it. When a component isn't activated, the items in it appear transparent on the work plane (Figure 3-7).

FIGURE 3-6: Make a component using one of these two methods to ensure all parametric history is stored inside it.

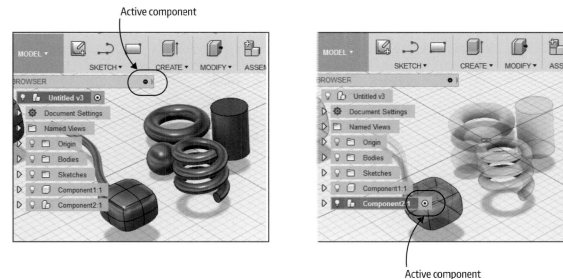

Active component

Active component

FIGURE 3-7: When a component is active, it will appear dark. When it's inactive, it will appear transparent.

MAKING BODIES WITH CONSTRUCTION TOOLS

Let's make some bodies now with the Model workspace's construction tools: Extrude, Revolve, Sweep, Loft, Rib, and Web. They're all under the Create menu. Using the dialog box that appears while using the tool, you can choose from options that include Join, Cut, and Intersect. Join welds bodies together, Cut subtracts one intersecting body from another, and Intersect leaves the overlapping parts of multiple bodies and deletes everything else.

Extrude

Extrude turns a sketch into a body. When you extrude a sketch, a handle appears with which you can taper the top plane (Figure 3-8). This handle doesn't appear in future extrude operations on that body. The Extrude tool also cuts through a body (Figure 3-9). To extrude multiple sketches the same height, extrude one at the desired height, and before clicking the OK button, click the other sketches. They'll extrude to that same height.

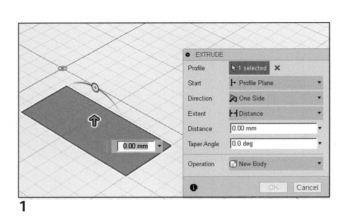

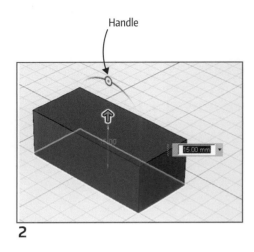

Handle

1

2

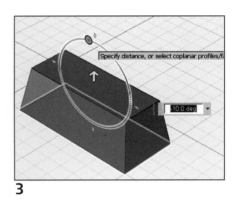

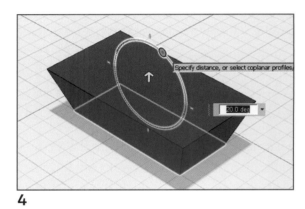

3

4

FIGURE 3-8: The Extrude tool turns a sketch into a body. A handle appears when you first extrude the sketch that you can use to taper the top plane.

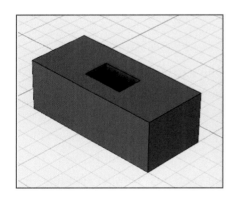

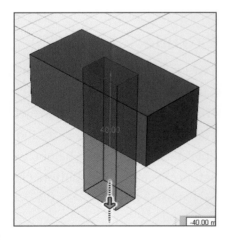

FIGURE 3-9: The Extrude tool cuts through a body.

Sweep

Sweep extrudes a sketch along a path (Figure 3-10). Select the sketch to sweep and then select the path to sweep along. If there are multiple sketch paths, you must select them in sequence. Then check the Chain box so that they'll all get swept.

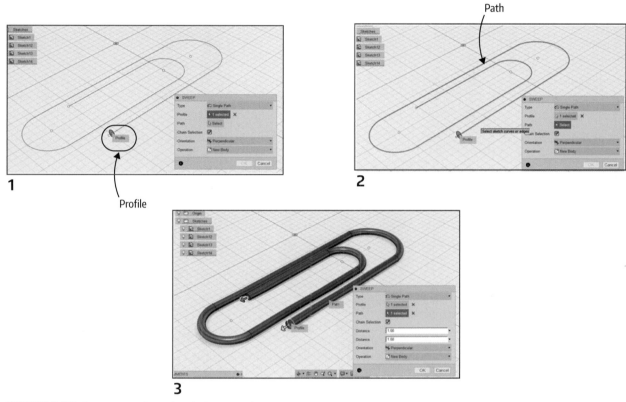

FIGURE 3-10: Sweep extrudes a sketch along a path.

Revolve

Revolve rotates a sketch around an axis (Figure 3-11). Select the sketch to revolve and then select the axis to revolve it around. The space relationship between the axis and sketch is important. For example, if you're making a chess piece, the axis should be in the center of the piece. If you're making a doughnut, there should be space between the sketch and the axis it is revolved around.

Profile sketch

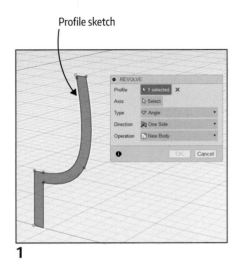

1

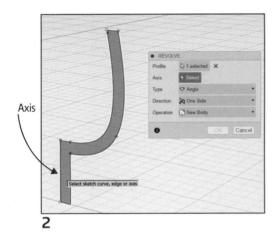

Axis

2

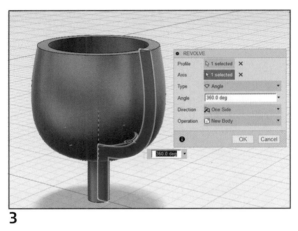

3

FIGURE 3-11: Revolve rotates a sketch around an axis.

Loft

Loft interpolates a body between sketches. Select the sketches in the order you want to loft them. Click each one to select—don't hold the Shift key while selecting, and don't drag a selection window around the sketches. After they're selected, activate the Loft tool. Alternatively, you can also loft two sketches at a time, which may give different results. You can loft to a point instead of a sketch and loft single-line sketches (Figure 3-12).

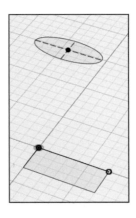

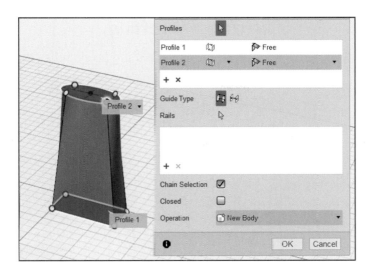

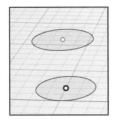

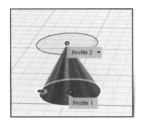

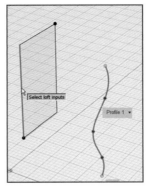

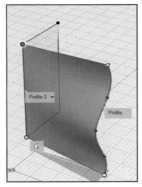

FIGURE 3-12: The Loft tool interpolates a form between sketches or points.

Sketching for Lofting

Here are some tips for lining up sketches that you want to loft. First draw a "skeleton" of lines on the vertical plane. Place the horizontal lines the distance apart that you want the sketches to be. In the Construct menu, click Plane At Angle (Figure 3-13).

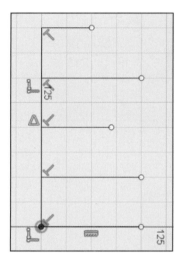
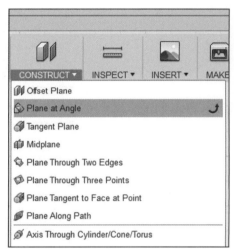

FIGURE 3-13: Draw a skeleton of lines on the vertical plane and then click Construct → Plane At Angle.

Select a horizontal line. A plane will appear on that line; either drag the arrow or type the angle you want it to be, and then press Enter. Repeat on each line. Then sketch on each plane (Figure 3-14).

After you loft all the sketches, you can tweak the loft's shape by selecting one of the skeleton lines and moving it. The sketch plane will move with the line (Figure 3-15), which changes the loft's shape.

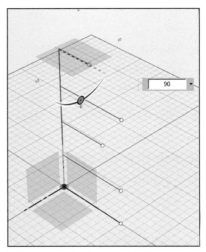
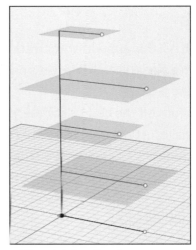
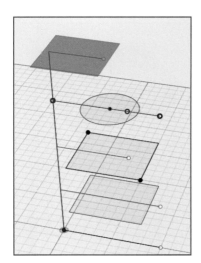

FIGURE 3-14: Place sketch planes on each line and then sketch on the planes.

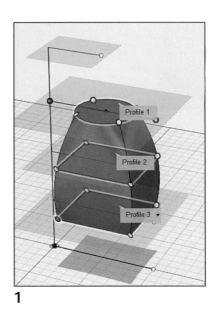

1

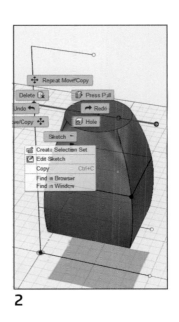

2

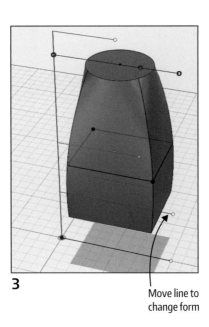

3

Move line to
change form

FIGURE 3-15: Change the loft shape by moving the lines the sketch planes are on.

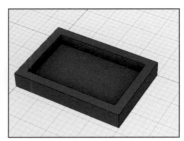

FIGURE 3-16: The box in which we'll put a diagonal rib

Rib

A rib is a thin support that creates rigidity. The Rib tool creates one from a single sketch line. Let's put a diagonal rib in the box shown in Figure 3-16.

Turn off the body (click the light bulb in front of its Browser entry) to make just the sketch visible. Then draw a diagonal line. Activate the Rib tool, click somewhere on the line, and drag it up. You can type specific thickness and depth in the text fields (Figure 3-17).

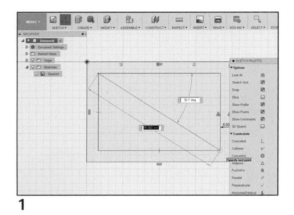

1

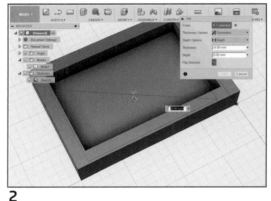

2

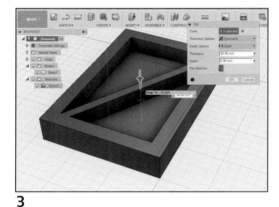

3

FIGURE 3-17: Modeling a rib

Web

The Web tool creates multiple ribs—that is, multiple thin features from single sketch lines. First, draw the lines. Then activate the Web tool and select the web sketch lines (Figure 3-18). The sketch lines don't have to reach all the way to the solids or to any other geometry, because when the ribs are created, they will automatically extend to all boundaries. If you don't want the web to extend to the boundaries, offset planes from those boundaries, and the web will extend to the planes instead (Figure 3-19).

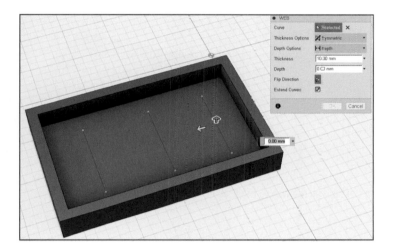

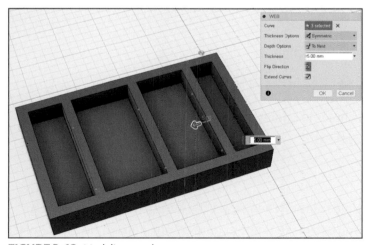

FIGURE 3-18: Modeling a web

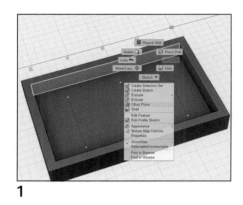

1

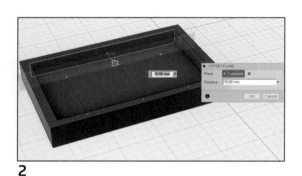

2

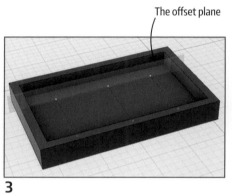

The offset plane

3

FIGURE 3-19: Offset planes so the rib or web doesn't extend all the way to the solid boundaries.

> **TIP** If the Rib or Web tool doesn't produce the desired results, try different settings in the tool's dialog box. For example, select Flip Direction for the Web tool or Depth Options for the Rib tool.

HOLES AND THREADS

You can create holes in bodies and threads out of cylinders. Each tool has multiple options.

Hole

Activate the Hole tool and then select the face on which you want to make a hole. You can make simple, counterbore, and countersink holes, and adjust the tip angle (Figure 3-20). Drag the center circle to place the hole, and drag the handles to adjust the diameter.

Handles   Center circle

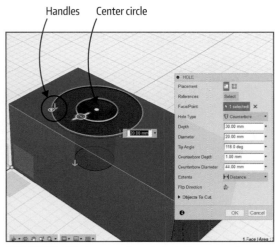

FIGURE 3-20: A counterbore hole being created

Thread

These cuts are made on cylinders. Activate the Thread tool, select a face, and then choose the options (Figure 3-21). The cylinder can be inside a solid, such as the hex nut in Figure 3-22. Check the Modeled box to make the threads permanent. The threads won't export in an STL file without checking that box.

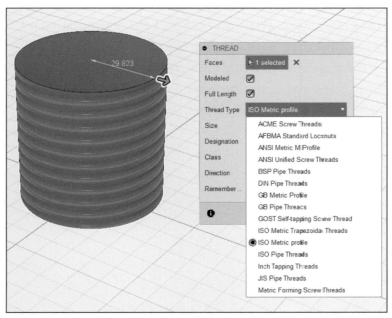

FIGURE 3-21: Modeling threads on a cylinder

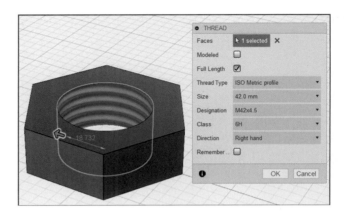

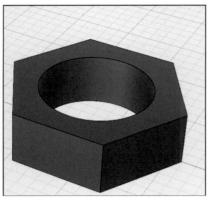

FIGURE 3-22: Modeling threads inside a hex nut

Sculpt icon

Create form

FIGURE 3-23: The Create Form icon and menu option

THE SCULPT WORKSPACE

Click the Create Form icon to enter the Sculpt workspace. If the icon doesn't appear on your toolbar, scroll to the bottom of the Create menu and choose it (Figure 3-23).

Make a Sculpt Box

The tool menu changes when you enter the Sculpt workspace and the Create menu is shorter (Figure 3-24). This workspace makes curvy, organic-shaped forms, not linear ones. Let's make a box to demonstrate. In the Create menu, click Box. Then click the horizontal plane, click twice to sketch a rectangle, and extrude the rectangle up (Figure 3-25). If there are no lines on your box (as there are in Figure 3-25), click the Display Settings icon at the bottom of the screen and choose Visual Style → Shaded With Visible Edges Only.

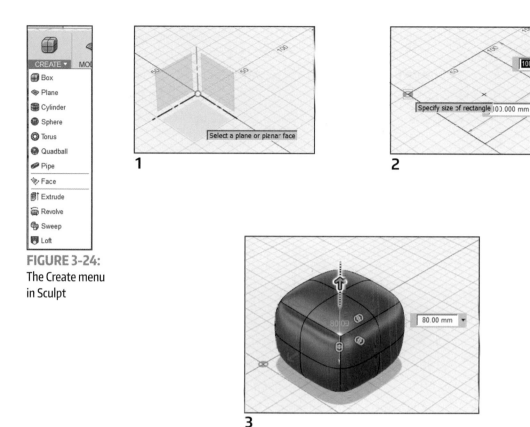

FIGURE 3-24:
The Create menu
in Sculpt

FIGURE 3-25: Modeling a Sculpt box

Plane and Quadball

These T-spline shapes are useful as a base to sculpt on.

PLANE This is just a flat surface on which you can sculpt. Click Plane, drag the arrows to set the size, and then drag the handles to set the number of faces (Figure 3-26).

QUADBALL This is similar to a sphere, but it is subdivided with quads, which are four-sided polygons. These enable different editing and are typically easier to work with and make better topologies. Quads stretch from edge to edge more cleanly than polygons (polygons stretch from corner to corner more cleanly). Click the tool and a quadball appears; drag the arrow or type a dimension (Figure 3-27).

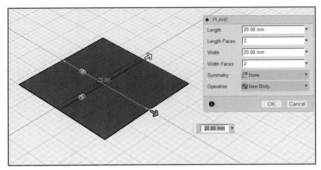

1 Size the plane

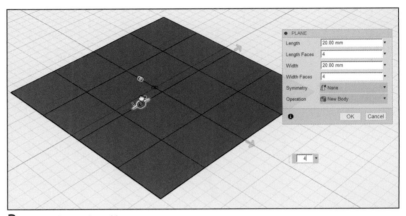

2 Choose the number of faces

FIGURE 3-26: Making a plane

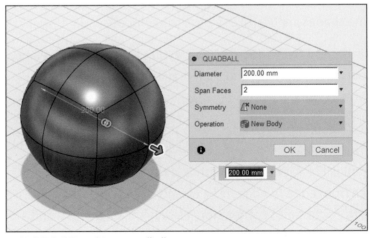

FIGURE 3-27: Making a quadball

Pipe and Thicken

There's a Pipe tool in the Model workspace that we've already discussed, but let's make one here to see how it's different. Make a body, select a path on it (it can be a sketch or body edge), choose options in the dialog box as desired, and press Enter (Figure 3-28).

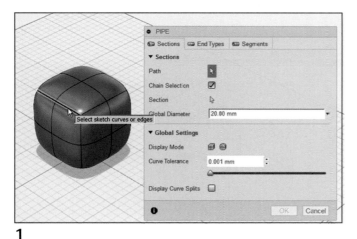

1

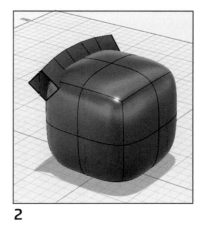

2

Pipe on an edge

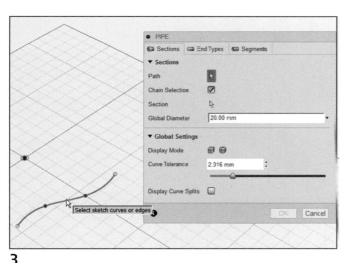

3

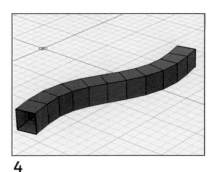

4

Pipe on a spline

FIGURE 3-28: Paths used to make pipes in the Sculpt workspace

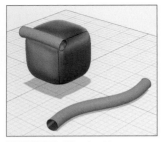

FIGURE 3-29: The Pipe command along a body edge and along a line

These pipes have no thickness. Bring them into either the Model or the Patch workspace and they'll appear yellow (Figure 3-29). This color indicates no thickness and must be fixed to be 3D printed. Select the pipes, right-click, and choose Thicken. Then drag an appropriate thickness with the arrow or type a number (Figure 3-30). It's best to model thickness as a parametric feature, not in the Sculpt workspace, for later editability.

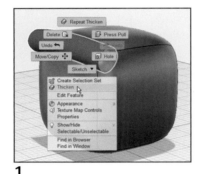

1

2

Thicken the edge pipe.

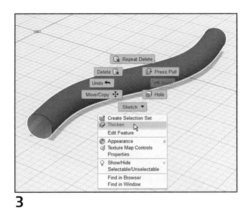

3

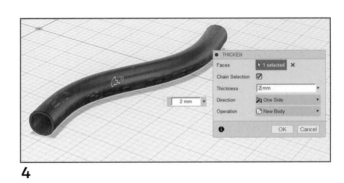

4

Thicken the spline pipe.

FIGURE 3-30: Thickening a pipe

Extrude and Loft

Back in Sculpt mode, let's see how its Extrude and Loft tools work. Make a form in Sculpt (I made a cylinder). Select a sketch, profile, or face and then drag the arrow and tweak the handle (Figure 3-31).

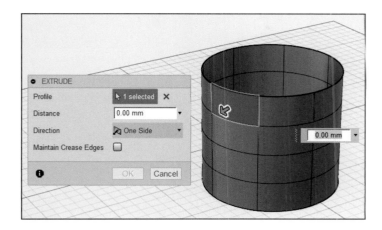

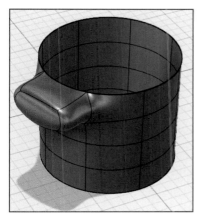

FIGURE 3-31: Extruding a face on a Sculpt cylinder

Next, make some sketches and loft them (Figure 3-32).

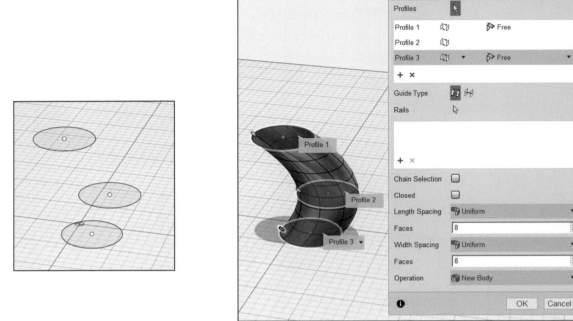

FIGURE 3-32: Sketches drawn and lofted in the Sculpt workspace

THE PATCH WORKSPACE

The Patch workspace makes models that have crisp, angular lines as in the Model workspace, but they're surface models, not solid ones. You can delete their faces by selecting them and pressing Delete or by right-clicking and choosing Delete (Figure 3-33). A workflow involves bringing a body made in the Model workspace—where you can't delete faces—into the Patch workspace, where you can. Figure 3-34 shows Patch's Create menu.

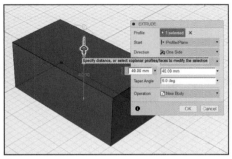

1 Make a box.

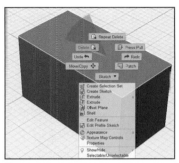

2 Select the top face.

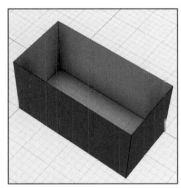

3 Delete the top face.

FIGURE 3-33: Delete faces in the Patch workspace.

Boundary Fill

The Boundary Fill tool is in both the Model and Patch workspaces. It creates, joins, or removes volumes. It's similar to the Combine tools (Join, Cut, Intersect), but works across multiple bodies and components, and provides options for what to do with the resultant geometry. This is a complex tool; we'll do a simple example here. The component that the body is in must be active, and results are easier to see if you change the model's appearance to something besides the steel material default. A ceramic appearance was applied in this example. We discuss changing appearance in Chapter 4.

FIGURE 3-34: The Create menu in Patch

1. Select the bottom plane, right-click, and choose Offset Plane (Figure 3-35).

2. Drag a construction window around the box and plane to select them (Figure 3-36).

3. Click each cell to preview what it will look like and choose the one you want. Here I chose the middle cell. Figure 3-37 shows the new volume made inside the boundaries of the box and offset plane.

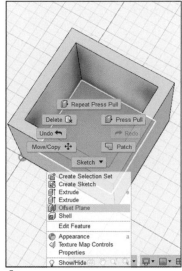

A Offset a plane from the bottom face

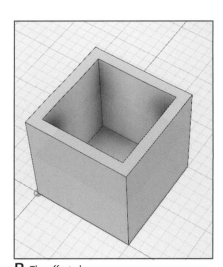

B The offset plane

FIGURE 3-35: Offset a plane from the bottom face.

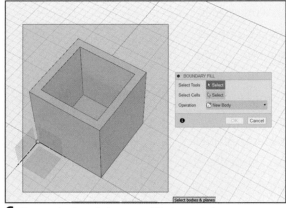

C Drag a selection window around the box and plane

FIGURE 3-36: Select the box and plane.

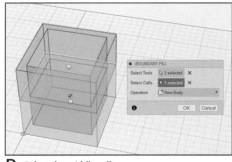

D Select the middle cell

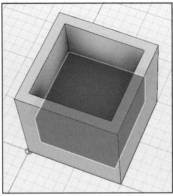

E The new volume made inside the boundaries of the box and offset plane

FIGURE 3-37: The new boundary volume

Offset

This Patch tool creates a new surface a distance you choose from a face.

Select the face to offset and then specify the distance (Figure 3-38).

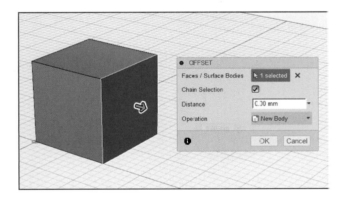

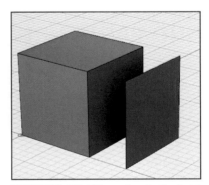

FIGURE 3-38: Offset a face in the Patch workspace.

Patch a Hole

Figure 3-39 shows a box made in Sculpt, with its top selected and deleted. Bring it into the Patch workspace, click Create ➜ Patch, and then click the edges of the surface to cover. A flat surface will cover the opening (Figure 3-40).

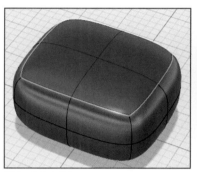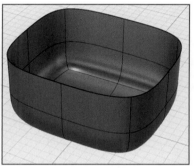

FIGURE 3-39: Make multiple selections by pressing and holding Shift, and then press Delete.

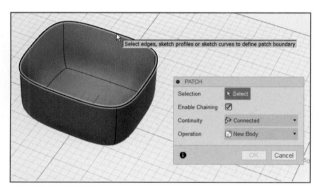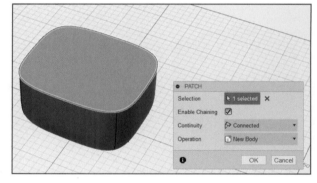

FIGURE 3-40: Covering a hole

COPY, MOVE, PASTE, AND PASTE NEW

The methods described here work in both direct and parametric modeling mode.

Copy

To copy a body or component (we'll refer to both as a part), select it by either dragging a window on it or clicking its Browser entry.

Right-click and choose Move/Copy. In the resulting dialog box, select the Create Copy box; by default it is unchecked (Figure 3-41). A widget will appear; use it to drag the copy off the original (Figure 3-42). Bodies can be copied and pasted inside components. When a component is parametrically complete, all copies of it will also be parametrically complete.

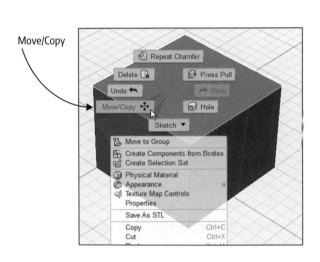

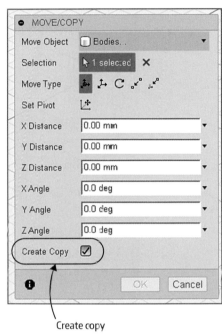

FIGURE 3-41: Select the item, right-click, choose Move/Copy, and select the Create Copy box.

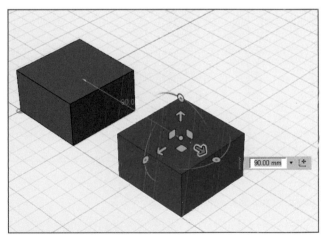

FIGURE 3-42: Drag the copy off the original with the transformer.

Move

If you just want to move the part, leave the Create Copy box dese-lected. Sometimes you can't move a part by directly selecting it; you must select it through the Browser, and then the widget will appear.

Copying a component is a bit more involved if you want to preserve the component's ability to automatically update all copies made of it. Right-click the Browser entry of the component that you want to copy and select Copy. It doesn't matter if the component is active or not when you do this. However, if it's active, deactivate it by clicking a different component after you choose Copy. If you don't activate a different component, you'll get an error message saying, "The selection cannot be copied because it contains the active component."

Paste

In Figure 3-43 I activated the root component and then right-clicked that root component and chose Paste. When you do this, a widget appears that you can use to drag the copy—called an *instance*—off the original. Pasted copies are linked to the originals so that when you edit the original or any instance of it, all instances will update. Linked com-ponents have a chain graphic next to their Browser entry.

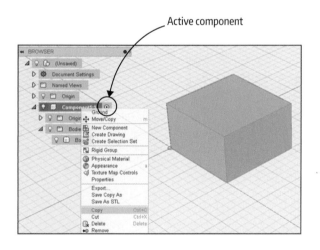

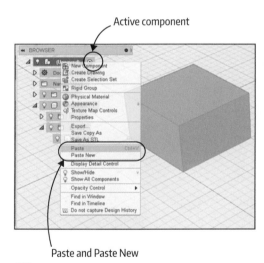

FIGURE 3-43: Right-click the component to copy and then paste it into a different component.

If you edit a component and the other instances don't edit, you might have created a body instead of a component, you might have right-clicked a body entry in the Browser, or the component might not have been activated when you dragged a body into it. The best way to make a component whose instances will update with edits is to right-click the Browser root, choose Component, and then make bodies inside it.

Paste New

If you don't want a specific instance to update when changing the original, choose the Paste New option from the context menu instead of Paste.

MIRROR BODIES AND COMPONENTS

Bodies and components are mirrored the same in both direct and parametric mode. Click Create ➜ Mirror and then select the item. Then click the Select button in the dialog box. The origin planes will appear. Click the plane you want to mirror around (you don't have to mirror on an origin plane; you can create your own). A preview will appear; click OK if it's what you want (Figure 3-44). To move the mirrored copy, right-click its Browser entry and drag the copy off the original (Figure 3-45).

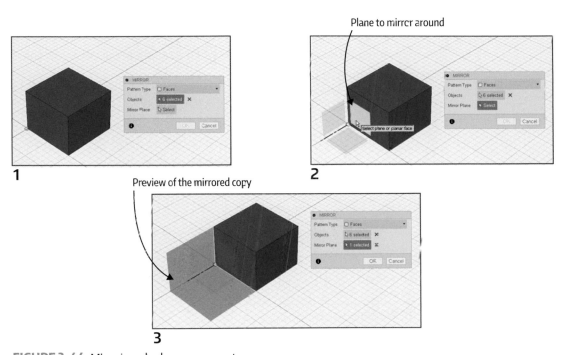

FIGURE 3-44: Mirroring a body or component

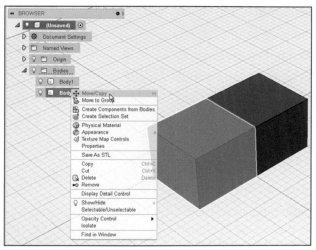
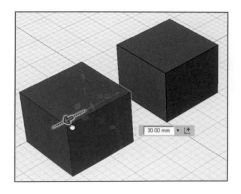

FIGURE 3-45: Moving the mirrored copy

PATTERN BODIES AND COMPONENTS

The Pattern tool has three options (Figure 3-46). The first two, Rectangular and Circular, work similarly to what was covered in Chapter 2, "Sketching." Pattern On Path copies and arrays a face, feature, body, or component along a specified track. Select the part and then select the path (Figure 3-47). The part doesn't have to be on the path; it can be anywhere on the work screen and will make and array copies in the direction of the path.

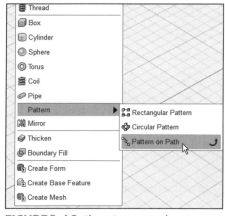

FIGURE 3-46: Three Pattern tool options

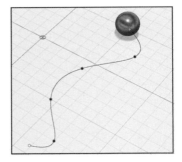

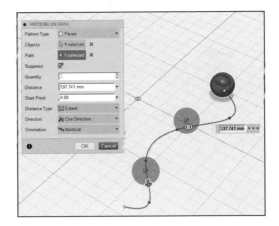

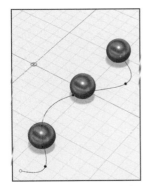

FIGURE 3-47: Using the Pattern tool to make copies of a sphere along a spline path

THE CREATE BASE FEATURE

Three entries appear at the bottom of the Create menu. One is Create Form, which puts you into the Sculpt workspace, and another is Create Mesh, which lets you modify and repair mesh files.

The third is the Create Base feature, which puts you into direct modeling mode within the parametric modeling mode. This is great for features that are difficult to model parametrically and for which you don't need a timeline. No parameters are generated or used in a base feature, but you can add parametric features to base features later—for example, by sketching on a base feature face. When you're done editing, the parametric features will update.

So now that we've discussed modeling tools, join me in Chapter 4 and we'll explore the editing tools.

4

EDITING MODELS: THE MODIFY MENU

In this chapter, you'll learn how to use the Modify menu's editing tools. We'll work in Model, Sculpt, and Patch modes. Figure 4-1 shows the Modify menus in all three workspaces.

FIGURE 4-1: The Modify menus in the Model, Sculpt, and Patch workspaces

TOOLS IN THE MODEL WORKSPACE

Let's start by looking at the tools in the Model workspace's Modify menu.

Shell

The Shell tool hollows a form, leaving walls with a thickness that you specify. Select a face or faces, and then specify a thickness. The selected face(s) will be removed and the body will become hollow. Figure 4-2 shows two examples of shelling. This tool makes it easy to create things like piggy banks and boxes out of solid forms.

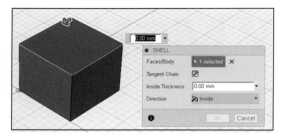
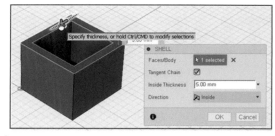

One face shelled

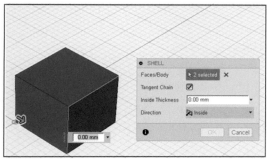

Two faces shelled

FIGURE 4-2: The Shell tool hollows a form.

Draft

Draft is a slope incorporated into a die-casting piece—a mold into which molten metal is poured. Those who use silicone to make molds of 3D-printed parts also work drafts into their designs. A draft angle helps remove the finished part from the mold or, in the case of sand casting, helps remove the model part before casting without destroying the sand mold.

First, select the neutral plane and then select the face you want to draft. Next, drag the handle to create the slope or type a number (Figure 4-3). The Draft tool applies a draft angle to planar faces.

Handle

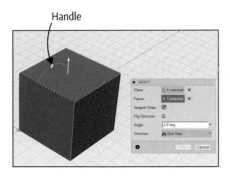
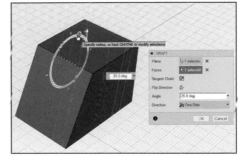

FIGURE 4-3: Making a draft angle

Scale

The Scale tool scales sketches, bodies, and components (Figure 4-4). Select the entity you want to scale and then drag the arrow or type a number. An arrow appears at the point the item will scale around; to change that point, click the Point button in the dialog box and click a new point. All sides will scale uniformly. To scale along one axis, click the dropdown arrow in the dialog box and choose Non-uniform. Three arrows appear; drag the arrow on the axis you want to scale along.

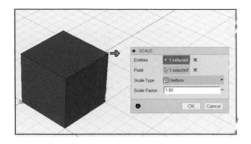
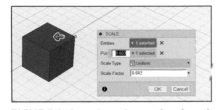
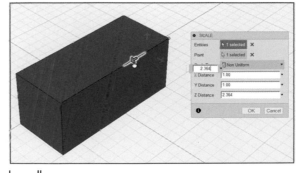

FIGURE 4-4: Scaling a box uniformly and non-uniformly along all axes

Scaling a Sphere

Spheres don't have a point to scale around, so you need to make one. Choose Construct ➔ Point At Center Of Circle/Sphere/Torus. Select the sphere and a point will appear at its center. Click OK. You'll now be able to use the Scale tool on it (Figure 4-5).

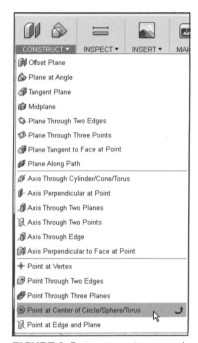

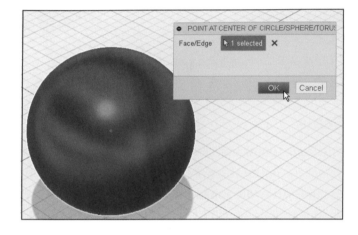

FIGURE 4-5: Create a point on a sphere to scale it.

Combine

The Combine tool joins, cuts, and intersects solid bodies (called Boolean operations). All options work the same way. You select the target body, which is the body which you want to change. Then you select one or more tool bodies to make that change with. The order of body selection is important, because you get different results with different orders. Combine → Join must be applied to a model that has multiple pieces to make it 3D-printable. Figures 4-6, 4-7, and 4-8 show these three operations performed on a box and cylinder.

Target   Tool

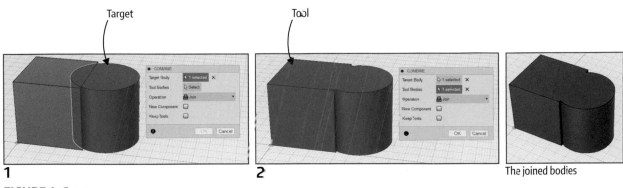

1   2   The joined bodies

FIGURE 4-6: Join

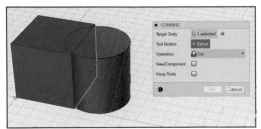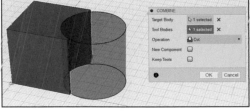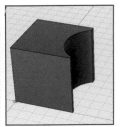

FIGURE 4-7: Cut

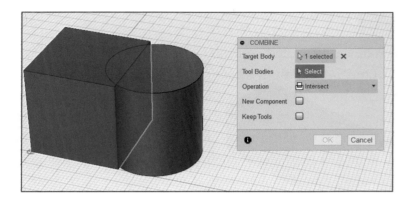

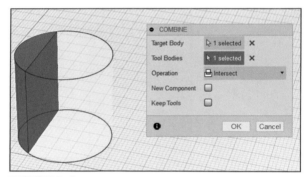

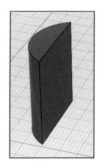

FIGURE 4-8: Intersect

Replace Face

A face is a surface on a body or component. The Replace Face operation replaces one face with another. First, select the face to remove (the target), and then select the replacement (source) face. The old face will disappear, and the surface will take the shape of the replacement face. After this operation, you can move the replacement face away (Figure 4-9) and delete it.

For this tool to work, the replacement face must be one continuous piece and completely intersect the old face. The examples shown in Figure 4-10 won't work. One was made from multiple arcs and thus has multiple edges, so it's not one continuous piece. The other is smaller than the replacement face.

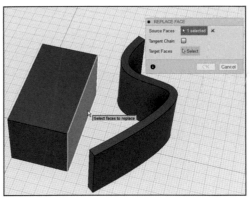

1 Select the face to replace.

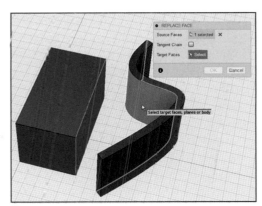

2 Select the replacement face.

3 The two bodies will snap together.

FIGURE 4-9: Replacing a face

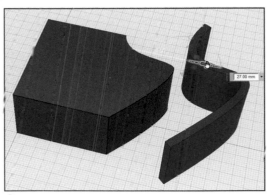

4 Move or delete the extra face.

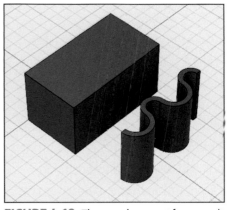

FIGURE 4-10: These replacement faces won't work. One is not continuous and the other is too small.

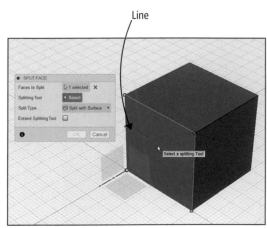

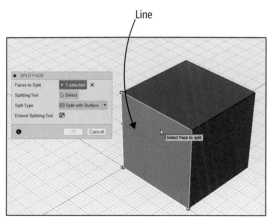

1 Select the face to split

2 Select the line

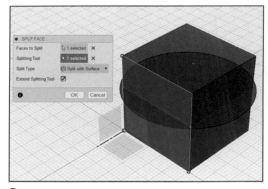

3 A preview of the split face

FIGURE 4-11: Splitting a face

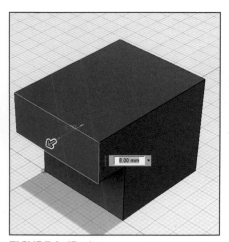

FIGURE 4-12: There are now two separate faces, and operations can be applied to each.

Split Face

The Split Face tool breaks a face into two faces. Sketch a line where you want to make the break. It doesn't matter if the line is sketched directly on the face or on another plane. Activate the tool, click the plane you want to split, click the blue Splitting Tool button, and click the line. A red surface will appear at the split location. Click OK (Figure 4-11). The face is now broken in two, and separate operations can be applied to both (Figure 4-12). Note that the Split Type option in the dialog box was set to Split With Surface.

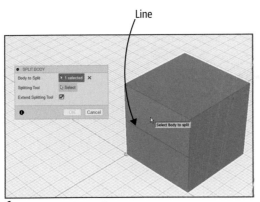

Line

1 Select the body to split

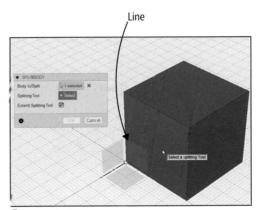

Line

2 Select the line

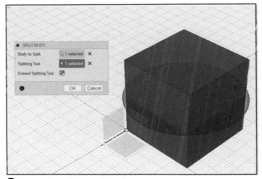

3 A preview of the split body

FIGURE 4-13: Splitting a body

Split Body

The Split Body tool splits a body into two bodies, especially useful for breaking a file into multiple parts for 3D printing. Sketch a line where you want to make the break. It doesn't matter if the line is sketched directly on the body or on another plane. Activate the Split Body tool, select the body you want to split, click the blue Splitting Tool button, and click the line. A red surface appears at the split location. Click OK (Figure 4-13). The body is now broken in two, and separate operations can be applied to them (Figure 4-14). You can also use an offset plane set anywhere along the body or a plane on another body for this purpose (Figure 4-15).

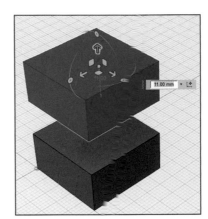

FIGURE 4-14: There are now two bodies and separate operations can be applied to them.

The offset plane

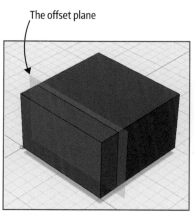

FIGURE 4-15: An offset plane can be placed anywhere along the body.

TIP When splitting a model into pieces for 3D printing, incorporate connectors into the design, which may avoid the need to glue the parts together.

Silhouette Split

The Silhouette Split tool splits a body in a manner that shows the body's entire outline. All points must be coplanar (on the same plane) for this tool to work. Click a plane to choose the viewing direction (Figure 4-16). Unlike with the Split Body tool, you can't choose a specific location to split. Next, click the body. It will split into two equal parts. This tool can also be applied to a face.

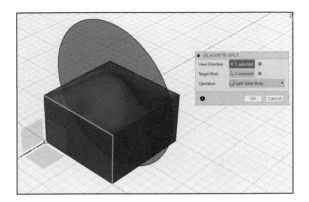
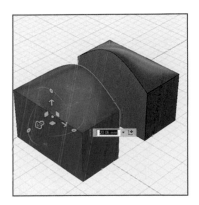

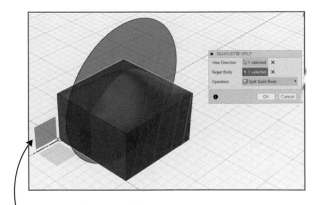

Choose plane in the viewing direction.

FIGURE 4-16: The Silhouette Split tool splits the body to show the body's entire cutline in a specified viewing direction.

Align

The Align tool snaps an entity (body, component sketch, construction geometry) to another by lining up geometry you pick on both of them. Select the point on the item you want to move, and then select the point on the face you want to align to (Figure 4-17). Click the Flip or Angle options to rotate the item.

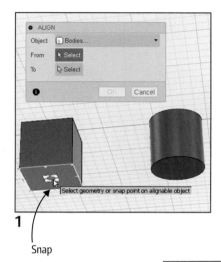

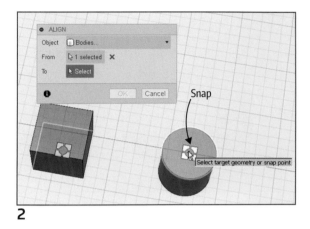

1

Snap

2

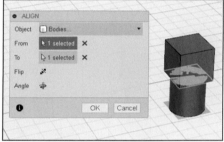

3 Box snaps to cylinder

FIGURE 4-17: Aligning a box to a cylinder

TIP When editing, it's best to roll back the timeline to that feature. Some features don't modify correctly unless you move the timeline back.

PHYSICAL MATERIALS AND APPEARANCE

The model has the default material of steel assigned to it. You can keep it as is or change to one of your liking, but you can't have a material-less model. Find or change it in Preferences → Material (Figure 4-18). Assigning appearances and materials has limited use to 3D printing and CNC construction because the color of your project will be the color of the real-life material. But these functions do have some uses that are worth a mention here.

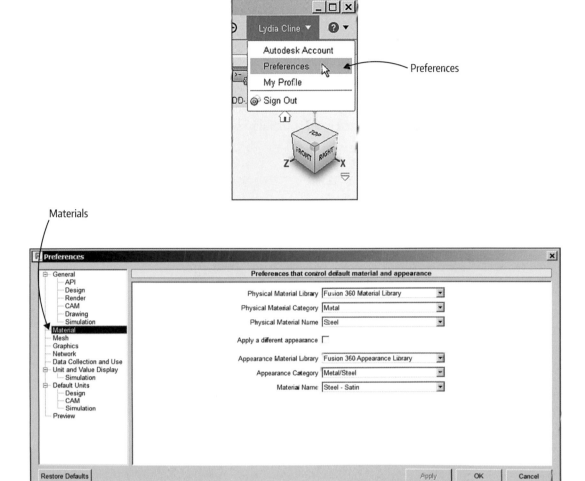

FIGURE 4-18: Preferences → Materials shows the default material.

Physical Materials

Physical materials affect the model's color and engineering properties (for example, mass and density) of bodies and components. Applying a color is relevant if you send your print to a commercial printing service that has machines capable of replicating colors. Drag the physical material's thumbnail from the dialog box to the model to apply it. To return to the default appearance, drag the default material icon back onto the model. All icons for materials in the model appear in the dialog box's In This Design field.

Appearance

Appearance affects the model's color, but not its engineering properties. Appearance overrides any colors assigned from the Physical
Materials function. The glass appearance is useful if you want your
project to look transparent so that you can inspect it for problems.
Drag the appearance from the dialog box to the body, component, or
face (Figure 4-19).

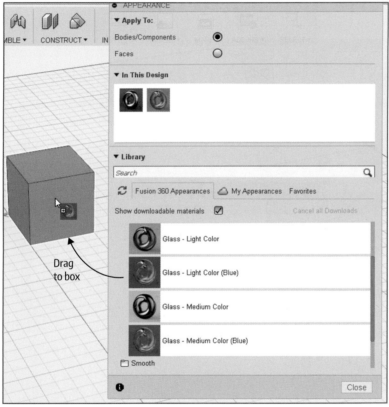
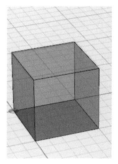

FIGURE 4-19: Apply a material by dragging it from its folder to the model. Here, the glass thumbnail is applied,
making the model transparent.

When you change the default material in Preferences, it affects subsequent designs but not existing ones. If a design clings to an appearance that you're trying to change, that appearance was probably applied to the top-level Browser field. Right-click that field, choose Appearance, and drag the desired appearance to the model (Figure 4-20). Remove all physical materials and appearances before exporting as an STL file, because they can conceal issues, such as smoothness, that you might want to address before export. Also, it's best to work with patterns and materials suppressed, because they slow Fusion down. You can suppress them, as you can any feature, by right-clicking their icons in the timeline and choosing Suppress Feature.

FIGURE 4-20: Change a model's appearance by accessing the Appearance function in the top-level Browser field.

DELETE, REMOVE, AND HIDE

Delete completely removes something from the timeline. Remove is a "soft" delete that removes something from the timeline going forward after the Remove function is applied. Hide makes a feature or entity invisible.

Delete

You can delete features, parts, and whole designs in one of three ways: using the Modify menu; deleting the timeline icon; or selecting, right-clicking, and then choosing Delete or pressing the Delete key (Figure 4-21).

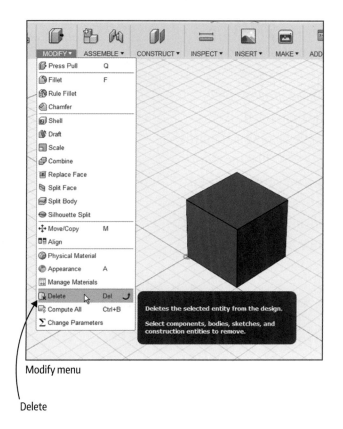

Modify menu

Delete

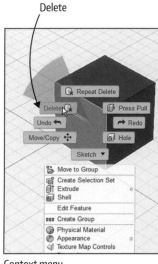

Context menu

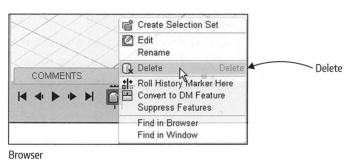

Browser

FIGURE 4-21: Three ways to delete features and whole models

If one way doesn't work, try another. For example, the point sketch that often appears at the origin point often can't be deleted by selecting and pressing the Delete key, but you may be able to delete it through its Browser entry (Figure 4-22) or by dragging a window around it and then deleting.

When in direct modeling mode, deleting is simple. Just select, right-click, and choose Delete. Recall that that to enter direct modeling mode, right-click the title field in the Browser and choose Do Not Capture Design History. Deletion in parametric modeling mode is usually more difficult. If you can't delete something, first check to see if it's entirely selected. Selecting it through the Browser or timeline will completely select it. If that doesn't delete it, maybe it can't be deleted; for example, in Model mode, you can't delete a face from a solid body.

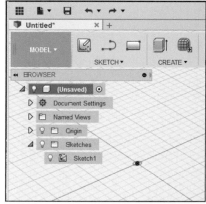

FIGURE 4-22: Delete an origin point sketch through the Browser.

You might be able to delete something but get a warning that the deletion will affect other features ("dependencies" in Fusion-speak). For example, suppose you applied a chamfer (angle) to four edges and now want to replace one of the edges with another feature. Look at the timeline. Yellow highlights will appear on the affected features. Fusion often caches enough information to enable it to compute a feature without a reference, enabling you to complete the deletion. But because the deletion might adversely affect something else, it's best to fix that yellow highlighted feature before continuing, such as redrawing the edge and then applying the new feature.

The deletion attempt might fail altogether, usually because it would mean a feature couldn't be calculated anymore. For example, maybe you're trying to delete an edge that a fillet (rounded corner) uses. The fillet won't intersect the model anymore. Or maybe you're trying to delete a sketch that a feature was built on. The affected feature will highlight red on the timeline. A solution to the scenario just described is to slide the timeline back to before the feature was made and create a new sketch. Then edit the red highlighted feature to tie it to the new sketch. Click the error message that appears for a detailed list of what's broken.

Generally, warnings and errors require you to:

▶ Remove a feature

▶ Edit a feature to supply the missing information

▶ Edit a feature so that the missing information isn't needed anymore

Whenever you reference geometry, you create the potential for failure if that geometry can't be found after an edit. There's actually a hierarchy of geometry for what's stable (harder to break):

1. Origin planes and axes (you can't break these)

2. Geometry created only from origin planes and axes (you can't break this, either)

3. Sketches (these can break—projected sketches are the hardest to fix)

4. BRep items (bodies, faces, edges, vertices)

Remove

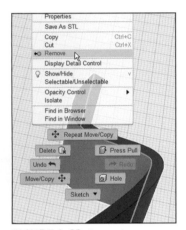

FIGURE 4-23: Removing a feature

If you can delete something, you may be able to remove it instead. Remove keeps an item in the timeline up until you apply the Remove tool. After you apply it, that feature is gone from that point forward. Right-click the selected item and choose Remove from its context menu (Figure 4-23). Only one body can be selected for removal at a time. The Remove tool won't appear if multiple bodies are selected.

Hide

Hide makes a model invisible. It's useful when you don't want to remove a feature but just get it temporarily out of the way. A reason to hide would be when you're exporting a model as separate STL files for dual-color printing. An STL file doesn't include hidden features.

To hide a feature, select it, right-click, and choose Hide. It will change to a lighter color; click in the workspace to make it disappear. To unhide it, find its entry in the Browser. It will have a gray light bulb in front of it; hovering the mouse over its entry will make it appear white. Click the gray light bulb to turn it back on (Figure 4-24).

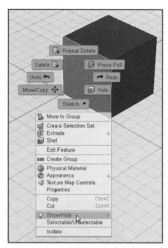
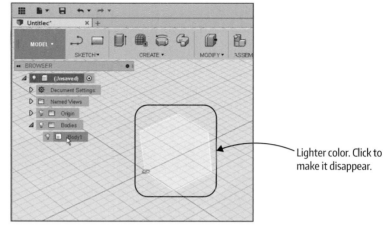

Lighter color. Click to make it disappear.

FIGURE 4-24: Hide a feature to remove it from view.

CHANGE PARAMETERS

Parameters are numbers, calculations, or equations you create to control sketches and 3D forms. At the very bottom of the Modify screen is the Change Parameters function (if you're doing parametric modeling). This lets you create parameters and apply them to sketches. Changing a parameter's value automatically changes all associated geometry.

Create a parameter first, and then add it to the dimension. To create one, click Change Parameters. In the dialog box that appears, click the plus sign to add a parameter. A parameter box appears in which you can type information. I named the parameter **Length** and kept the units set to mm. In the Expression box, enter a number, calculation, or equation. I typed **100** (Figure 4-25). Click OK, and the parameter will be listed under User Parameters.

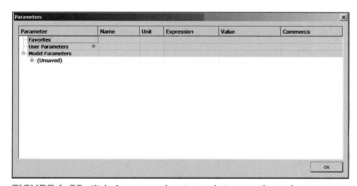

FIGURE 4-25: Click the green plus sign to bring up a box where you can name and describe parameters.

Now apply it to a sketch. I made the dimension text field of a sketch active by entering Edit Sketch mode, right-clicking a line, and clicking Sketch Dimension. I replaced the current dimension by typing **Length**, the name of the parameter. Note that it's case-sensitive. Press Enter, and the line adjusts to the parameter length and has a function symbol in front of it (Figure 4-26). You can do this only with sketches, not bodies and components.

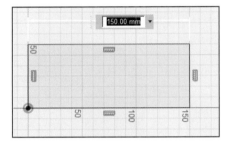 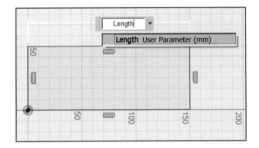

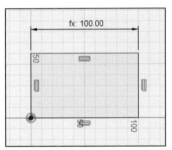

FIGURE 4-26: Applying the parameter to a sketch

To make a parameter reference another dimension, activate the sketch's dimension text fields, click the specific dimension you want to reference, and then add whatever equation you want to it.

DIRECT MODELING: EDIT FACE

Turn off the timeline to enter direct modeling mode. Now you can access two functions not available in parametric mode: Edit Face and Edit Feature (Figure 4-27). These tools will show up either in a context menu or under the Modify menu. They give you Sculpt-like capability while in Model mode. In Figure 4-28 I applied the Edit Face tool to the body of this tapered cylinder. The top face disappeared, and polygon edges appeared on the body. By clicking the vertices (edge intersections), I twisted it into another shape. Pressing Enter makes the top face reappear.

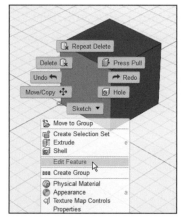

FIGURE 4-27: Access the Edit Face and Edit Feature tools via the Modify menu or a context menu.

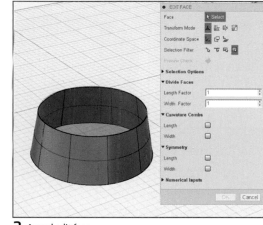

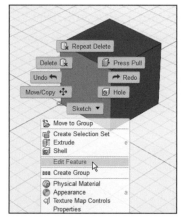

1 Select body.

2 Actual edit face

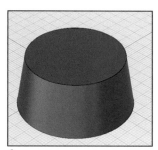

3 Trust the manipulators.

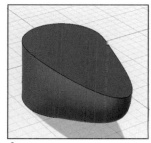

4 The result

FIGURE 4-28: Sculpting a form with the Edit Face tool

THE SCULPT WORKSPACE

Sculpt bodies are hollow. You can select and delete their faces (Figure 4-29), which you can't do on Model workspace bodies. Enter the Sculpt workspace and make a box with the Create ➜ Box tool. Select its faces and edges by clicking it. Double-click a face to select all faces connected to it; double-click an edge to select all edges connected to it. Let's apply some Modify tools to it now.

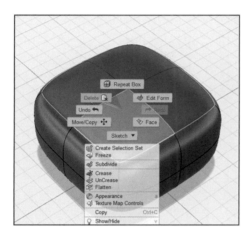 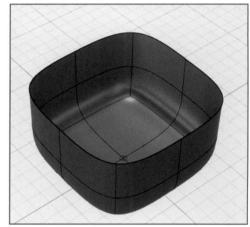

FIGURE 4-29: Faces on Sculpt bodies can be selected and deleted.

Subdivide Faces

Subdividing faces gives you more vertices to edit. Plus, it smooths the model out for 3D printing. Polygons appear in a 3D print, so on forms like spheres and curved corners, the more you have, the smoother those features will be. That said, you don't want to subdivide too much, because doing so will slow down the printer or make it impossible to print. Turn off any applied appearances or physical properties when subdividing for smoothness, because they hide the actual polygon count with an illusion of perfect smoothness.

Subdivide the faces of Sculpt bodies by selecting, right-clicking, and choosing Subdivide (Figure 4-30). Drag the vertices around to change the form (Figure 4-31). The more vertices, the more editing control you have. (If you don't see any lines and vertices, change the visual style under Display settings.)

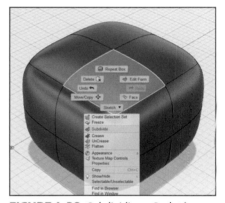

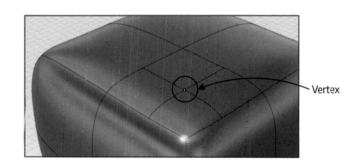

FIGURE 4-30: Subdividing a Sculpt box

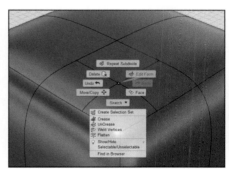

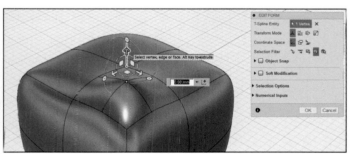

FIGURE 4-31: Drag vertices to change the body.

OTHER SCULPT TOOLS

Select some edges on a Sculpt box and right-click them to access context menu choices. Slide Edge moves the edges around (Figure 4-32).

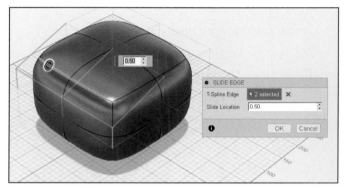
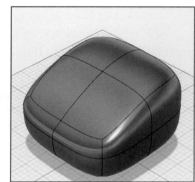

FIGURE 4-32: Slide Edge

Flatten repositions vertices (Figure 4-33).

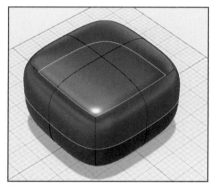
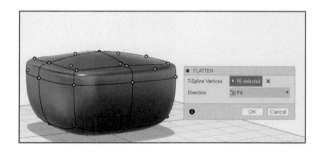

FIGURE 4-33: Flatten

Crease puts sharp edges between faces (Figure 4-34). All faces that are impacted by the crease operation turn yellow.

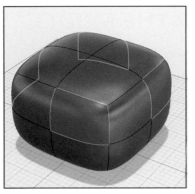
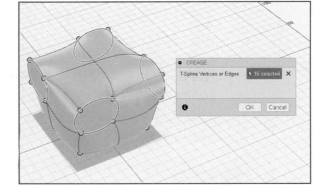

FIGURE 4-34: Crease

Bevel Edge replaces a single edge with a pair of edges, enabling you to modify sharpness along an edge. Figure 4-35 shows the effects of this.

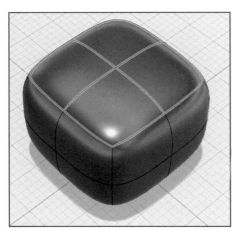

FIGURE 4-35: Bevel

Unweld Edges separates edges (Figure 4-36)—that is, it separates a body into an outer body and inner body. This is useful if you want to work on the outer body and not the inner body, or delete the inner body and re-thicken the outer body.

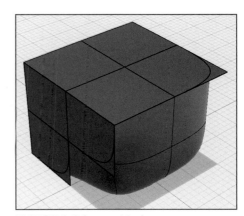

FIGURE 4-36: Unweld Edges

EDIT A SCULPT FILE IN THE MODEL WORKSPACE

When you're done with the Sculpt file, either click Finish Form on the toolbar or click Model on the toolbar menu. This puts you back into the Model workspace (Figure 4-37).

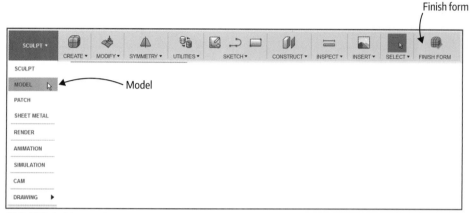

FIGURE 4-37: Return the Sculpt file to the Model workspace.

Thicken a Sculpt Box

Bring the box whose top faces we deleted earlier into the Model workspace (if you don't have it anymore, just make a new box and delete its top face). Then click the Thicken tool. An arrow will appear with which you can thicken the walls, which you must do to make them 3D-printable (Figure 4-38). You can thicken the box inside or outside its walls.

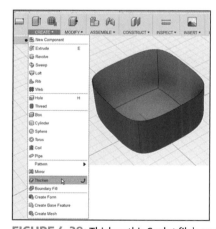
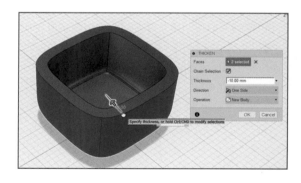

FIGURE 4-38: Thicken this Sculpt file's surface walls in the Model workspace.

Model a Puffy Bead

Let's turn that first box we made in the Sculpt workspace into a puffy bead in the Model workspace (turn on the timeline). Sketch a circle on it and extrude it through the box (Figure 4-39). Then select the circle's perimeter and right-click (Figure 4-40).

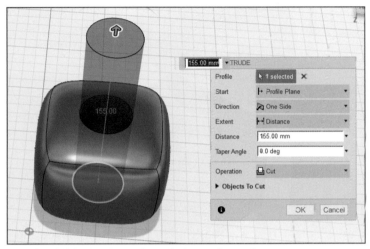

FIGURE 4-39: Sketch a circle on the box and extrude it.

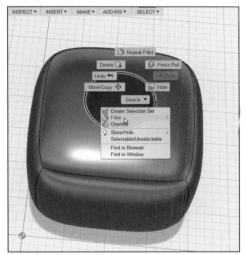

FIGURE 4-40: Select the circular edge and right-click.

Context menu options include Fillet, which gives a rounded edge, and Chamfer, which gives an angular edge (Figure 4-41).

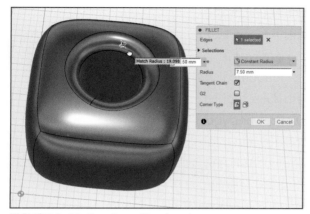
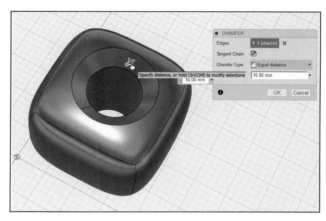

FIGURE 4-41: Chamfer or fillet the edge.

Edit Form

Figure 4-42 shows a plane that was made in the Sculpt workspace and subdivided with Sculpt's Subdivide tool. While in the Sculpt workspace, select one plane, right-click it, and choose Edit Form. A widget appears with which you can warp it.

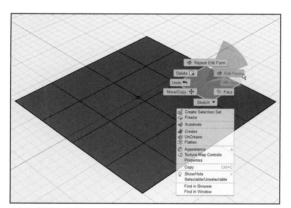
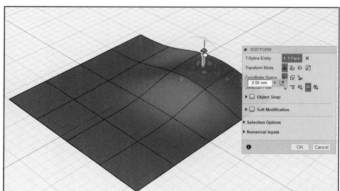

FIGURE 4-42: Warp a plane with the Edit Form tool.

THE PATCH WORKSPACE

Let's work with the Patch workspace's Modify menu now.

Edit a Funnel

Figure 4-43 shows a funnel made in the Model workspace by drawing
a profile with the Line tool and revolving that profile around its vertical
axis. Make this funnel and bring it into the Patch workspace. Delete
the top and bottom faces, then bring it back into Model and thicken its
walls (Figure 4-44). You could have made this funnel in Patch to begin
with, but there may be times that you want to use this technique on an
existing model.

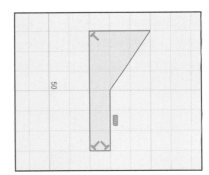

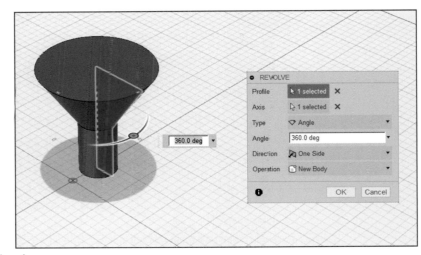

FIGURE 4-43: A funnel made in the Model workspace

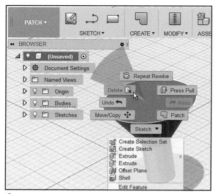

1 Delete the top face.

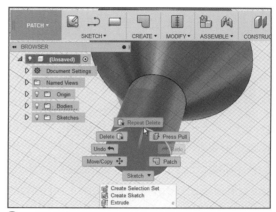

2 Delete the bottom face.

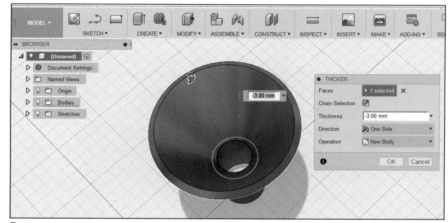

3 Thicken the walls.

FIGURE 4-44: Editing the funnel in both Patch and Model workspaces

Extend

The Extend tool makes a surface larger by extending it a distance you choose. Select the edge of the surface to extend and drag the arrow or type a distance (Figure 4-45). The extension type controls the direction of the new surface; the dropdown arrow in the dialog box next to Extend Type offers natural, perpendicular, and tangent options. (The surface in Figure 4-45 was made by lofting between two single-line sketches.)

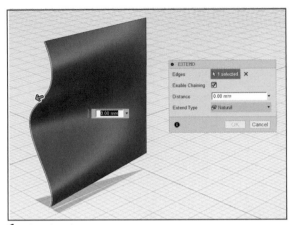

1 Select the edge.

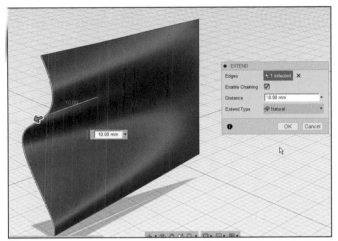

2 Pull.

FIGURE 4-45: Extending the edge of a surface

Trim

The Trim tool removes part of a surface body using a cutting tool. Figure 4-46 shows two intersecting surfaces. We want to trim one surface against another. Activate Trim. Click the surface to use as the Trim tool. Then click the surface you want to trim (Figure 4-47).

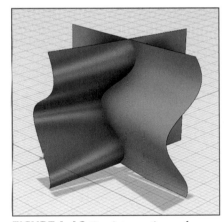

FIGURE 4-46: Two intersecting surfaces

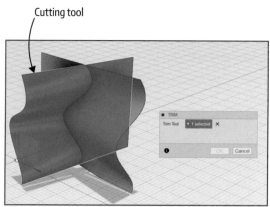

1 Select the surface to use as the Trim tool

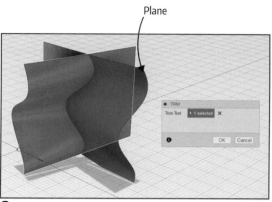

2 Select the plane to cut

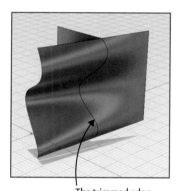

The trimmed edge

FIGURE 4-47: Trimming one surface against another

Reverse Normal

Faces have two sides, the back and front. The front is also called the normal and is a different color than the back. In Fusion, the normal is gray and the back is gold. All normals need to face out for a file to be

3D printable. If one or more backsides face out, click Reverse Normal, select the face the reverse, and click OK (Figure 4-48).

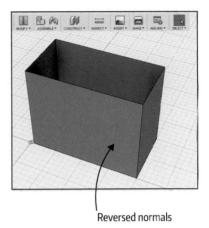

Reversed normals

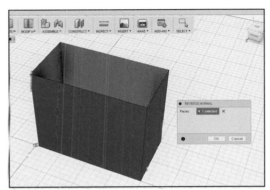

1 Select the faces to reverse.

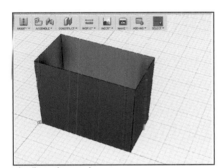

The flipped normals

FIGURE 4-48: Flip faces with the Reverse Normal tool.

Stitch and Unstitch

When you construct a model in Patch, the surfaces are stitched, or attached, by default. When you move it, attached faces will stretch along (Figure 4-49). To unattach a face, click Unstitch and select the face(s) to unstitch, and they'll become separate bodies, unattached from the rest of the model (Figure 4-50).

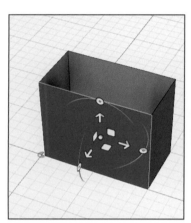
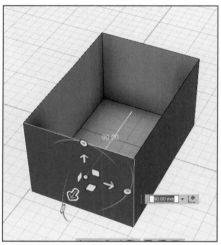

FIGURE 4-49: Stitched faces stretch when one is moved.

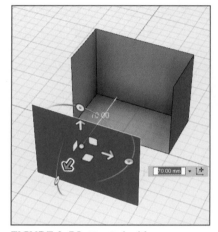

FIGURE 4-50: Unstitched faces can separate from the rest of the faces.

DEPENDENCIES

In parametric modeling, dependencies will play a large part in how you work. Dependencies are when one entity is dependent on another to exist. They cause issues like being unable to delete a file because a component in it is linked from another file, or being unable to move or scale a component because it is linked to another component.

In Figure 4-51, the business card file was opened first, and then the coin was dragged into that file from the Data panel. This created a dependency of the coin to the business card, shown by the link graphic in front of the coin entry in the Browser. You can move and scale the business card, but trying to move or scale the coin isn't possible; you won't be able to select it. Click the link to break it. That breaks the dependency, and you'll be able to edit the coin just like the card. If clicking it doesn't break it, right-click and choose Break Link from the context menu (Figure 4-52). You can relink a file by deleting it from the workspace, finding it in the Data panel, and dragging it back into the workspace.

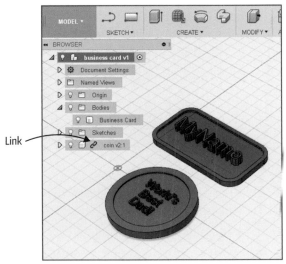

Link

FIGURE 4-51: When one body is dependent on another, it can't be freely edited.

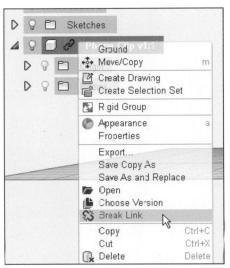

FIGURE 4-52: Break the link between bodies to edit each of them independently.

And that sums up the Modify menu tools! Join me in Chapter 5 now to learn best practices for modeling.

5

BEST PRACTICES AND NETFABB FOR FUSION

You can export any Fusion design as an STL. However, it will not necessarily print, or print well, even if it looks good on the screen. All designs should be optimized for best results. Optimization involves:

▶ Modeling the file well

▶ Orienting the file on the digital build plate for best printing results

▶ Running the file through analysis software to find and fix hidden defects

This chapter describes the characteristics of a digital model that will successfully print.

> **TIP** Rough modeling is common while going through the thought process of building a model. You may need to redo a rough model to make it printable. It often takes less time to redo a model than to try to fix a bad one. You'll model it faster the second time around, and it will be better crafted as you implement what you learned the first time.

BE WATERTIGHT

Watertight means the file has no defect holes in the mesh; if you poured liquid in it, none would leak out. If a model has holes, the printer won't know what to fill. If the model is supposed to have a hole, that hole must have an outside, inside, and defined thickness (Figure 5-1). This is not the same as a hole that is a defect in the mesh.

Water tightness is harder to achieve on polygonal models than solid models because just one missing edge or face will make it unprintable. Fusion's polygonal modeling mode, Sculpt, makes quads (4-sided polygons), which are easier to make 3D-printable than programs that make polygons with any number of sides ("ngons"). But you may have inserted a polygonal model made with another program into Fusion that needs repair even if your Sculpt file is fine.

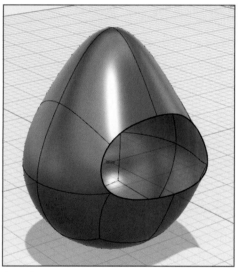 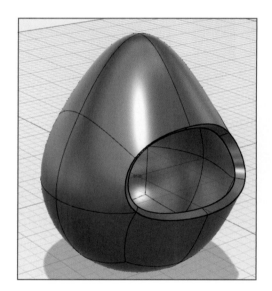

FIGURE 5-1: A hole must have thickness to be printable.

HAVE MANIFOLD EDGES

Each edge or vertex must connect to exactly two faces; these are called *manifold* edges. An edge that connects to three or more faces is called *non-manifold* and is unprintable. See Figure 5-2.

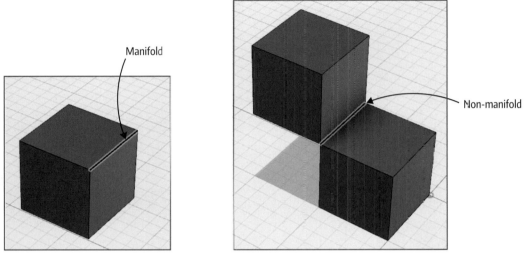

Manifold

Non-manifold

FIGURE 5-2: (Left) A manifold edge connects to two faces; (right) this non-manifold edge connects to four edges.

HAVE FRONT-FACING POLYGONS

Polygons have two faces, the front (also called the normal) and the back. The front must face out to be printable. Faces cannot overlap or have overlapping vertices. The back and front faces have different colors to tell them apart (Figure 5-3).

NO SURFACE INTERSECTIONS

The model must be one solid piece with no surface intersections. Otherwise, it will be unprintable or print with unexpected results. Figure 5-4 shows the model after the Combine tool has been applied to it, as well as a preview

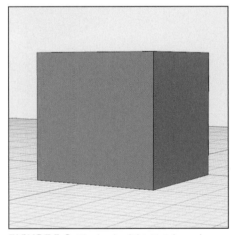

FIGURE 5-3: The back of this Patch mode model is facing out, as evidenced by the gold color.

showing that it will print well. Figure 5-5 shows the same, uncombined model and a preview of how it will print.

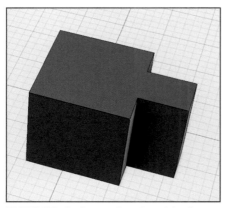

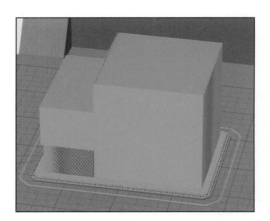

FIGURE 5-4: This combined model will print well.

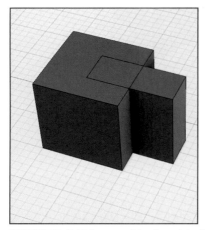

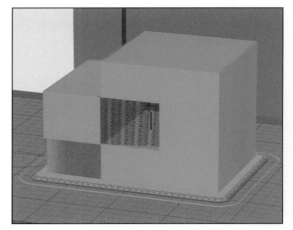

FIGURE 5-5: This uncombined model will not print as expected.

The Combine tool is necessary to weld parts together, but messy interiors may result. Nested geometry hidden inside other parts still gets printed, which can result in a longer print time, a failure partway through, or holes and other geometry problems. Shell the combined model, and then export it as an STL for printing (make sure your printer is set up to bridge that unsupported top). This will save on both material and print time.

For a clean result, polygonal models should have about the same number of subdivisions on the bodies to be combined. Polygonal models should also have fewer than 100,000 faces. A greater amount doesn't make the model measurably better looking; instead, it makes the model harder to work with. It may make combining bodies impossible. A high polygon count will also slow down the printer or make it impossible to print, especially if many of the faces are too tiny for the extruder.

HAVE APPROPRIATE CLEARANCES

Clearance is space that allows a good fit between interlocking parts. Leave enough clearance in a design that has attached parts such as links (Figure 5-6) so that parts don't print fused together.

Generally, clearances are between .25 mm and .75 mm depending on the size and shape of the model. Look for recommendations on the filament maker's website. ABS (acrylonitrile butadiene styrene) filament typically needs between .4 mm and .5 mm. On small items, start with a clearance of .15 mm. Exact clearances vary with the printer, slicer, and filament, so experiment with small parts before printing large ones. On large models, cut out the interlocking parts and print them to get the clearances right before printing the whole file.

Parts that fit snugly together, such as gears and cogs, should have a clearance between 0.1 mm and 0.25 mm built in on all sides to ensure they don't come apart easily after printing. Lay out the separate parts on the build plate to print (Figure 5-7).

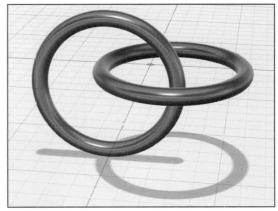

FIGURE 5-6: Leave enough clearance so parts don't fuse together when printed.

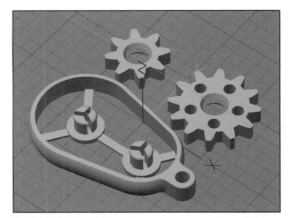

FIGURE 5-7: Build in a clearance on parts that will fit together after printing, and lay them out separately on the build plate.

HAVE APPROPRIATE WALL THICKNESSES

Walls need to be thick enough to be printed. They also have to be able to withstand removal from the build plate, removal of supports, post-processing, and shipping, if relevant. Supported walls should be at least 2 mm thick, and unsupported walls should be at least 3 mm thick (Figure 5-8). Thinner walls are likely to break, especially when supports are removed. This also applies to models that have a large mass connected to a thin one. Thicken pointed features. Files that are scaled down in the slicer to fit on a build plate may need to be returned to the original modeling software for arbitrary thickening.

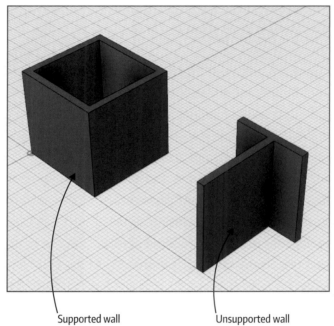

Supported wall   Unsupported wall

FIGURE 5-8: Walls should be the minimum printable thickness.

Digital models are often scaled down before printing, but a part that is scaled down too much won't print. Features that start thin or small get thinner or smaller and must be arbitrarily thickened or removed. Decisions must be made about what features to show and how to show them. This is especially important when using an online service bureau for your print, because assumptions will be made there on what to change, and those assumptions may be wrong. So always be cognizant of the design's suitability for printing if a 3D print is your end goal.

CONSIDER PLASTIC SHRINKAGE

Printed plastic changes dimensions when cooling; outer features expand and inner features shrink (Figure 5-9). To create dimensionally accurate parts, adjust the digital model to account for shrinkage. ABS filament shrinks about 2 percent and PLA shrinks about 0.2 percent. Hence, PLA is a better choice when dimensional accuracy is important. All features should be at least 2 mm in size.

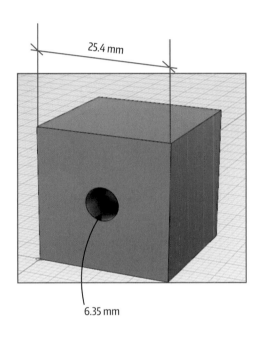

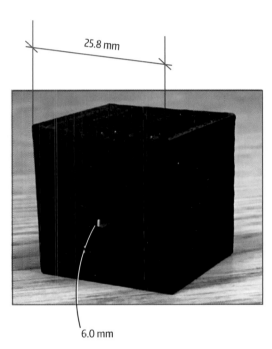

FIGURE 5-9: The print's size will vary from the digital model's size.

USE MILLIMETERS AS UNITS

You can design in imperial units if you're more comfortable with them, but export the model in millimeters, because that's the unit 3D printers understand. STL files don't contain units, just numbers, so your digital file may need some adjusting inside the slicer. Exporting the file in millimeters will enable you to easily scale the model inside the slicer.

HAVE GOOD ORIENTATION

The following are suggestions for orientation. Some may conflict with others, depending on the model. You may need to ignore one suggestion to implement another.

Orient to Avoid or Minimize Supports

Supports are structures that hold up overhangs (parts with empty space below). Slicing software generates supports, but you can finesse by adding, deleting, or moving them to other locations. After the print is finished, remove supports by snapping them off, cutting them off with a craft knife, or dissolving them, if they were made with dissolvable filament.

Supports are often difficult to remove, or fail during the printing process, so try to orient the model to avoid or minimize them. Features that are at a 45° or greater angle don't need supports, because each layer is built onto the one underneath it (Figure 5-10). Simply rotating the model can also cut down or eliminate the number of supports needed (Figure 5-11). You may be able to incorporate supports in the design, or slice the design in half with the Split Body tool to allow each part to lie flat on the build plate, and then glue the printed parts together.

FIGURE 5-10: (Left) The 45° features don't need supports; (right) the perpendicular overhangs need supports.

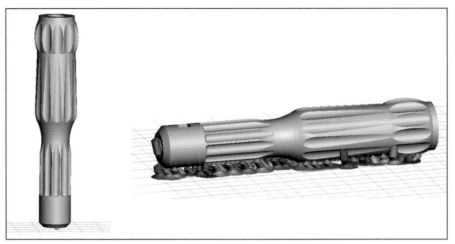

FIGURE 5-11: (Left) The screwdriver handle oriented straight up doesn't need any support; (right) the screwdriver handle oriented horizontally requires a lot of support.

Orient for Strength

Print items that need to be strong laterally on the Z layer, not vertically across multiple layers. For example, if you print the screwdriver handle in Figure 5-12 vertically, it will easily snap at each layer. But printed horizontally, the layers span the entire length, making the handle stronger.

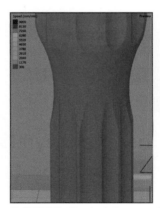

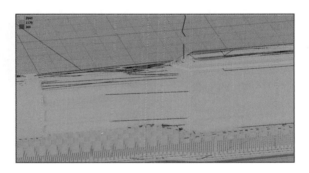

FIGURE 5-12: The left print will snap easily. The right print is oriented to be stronger.

Orient to Minimize Warping

The build plate is most level at the center, so parts printed there will warp less than parts printed closer to the edges. Heated build plates are also hotter at the center. Cooler print locations contribute to warping.

Orient to Avoid Stair-Stepping

Print sloped and curved surfaces horizontally (Figure 5-13). Printing them vertically results in a stair-stepping effect, especially when the surface has a shallow angle.

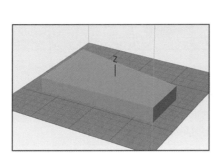
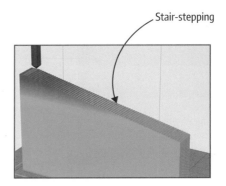

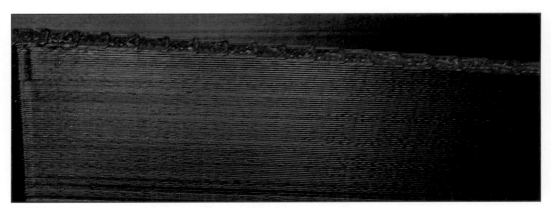

FIGURE 5-13: Stair-stepping on a vertically printed model

Orient to Preserve Detail

Thin layers print better vertically than horizontally. So position thin, detailed models like lithophanes (3D printed photos) vertically, as shown in Figure 5-14. On a printer with a moving bed, layers close to the build plate print best because they sway less than layers high up. This means that highly detailed features print better if they're oriented closer to the build plate. The mane on the lion in Figure 5-15 will print best if close to the build plate as shown.

FIGURE 5-14: Position a lithophane vertically for best detail printing.

FIGURE 5-15: Position detailed features closest to the build plate for best detail printing.

Supports often leave scars when removed, so orient a file that has detail you don't want ruined in a manner that avoids supports resting on them. Figure 5-16 shows how tilting the file 45° eliminates most of the supports that would otherwise mar the detail.

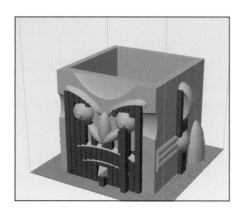

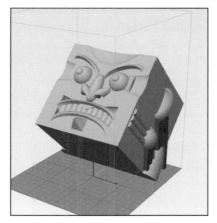

FIGURE 5-16: Orient the file to avoid putting supports on details.

Orient for Smoothness

A glass build plate gives the bottoms of prints a very smooth finish. Most of the time, this nice finish is wasted on the bottom of the print's base. If you have a file with a long outside face, such as a phone cover, you might want to orient it so it rests directly on the glass.

TEXT

Text on 3D prints can be either embossed (raised) or engraved (carved). Simple, blocky fonts are best because serifs and other flourishes are typically too small to successfully print. Engraved font prints best unless the width of an embossed font is thicker than 4 mm. Like all other features of the design, text should have the level of detail that the extruder is capable of printing.

INFILL

Infill is the density of the print. Thin, delicate prints need a high infill, such as 75–100 percent, or they will break easily. But 100 percent infill is not good for thick prints because of heat buildup during the printing process. In such cases, incorporate small holes into the design to let hot air escape. Most thick prints are fine with a 10–15 percent infill. A working part that will be put under stress may need a 40 percent infill and perhaps two or three shells (perimeter outline). Figure 5-17 shows the difference between a 10 percent and a 50 percent infill. A thin infill on a thin form will also result in a bendable print, which may be desired.

Design any needed holes in the digital model instead of drilling them into the printed part, because the drilling, and any nuts tightened onto the holes, can collapse the latticed infill. Honeycomb infills warp less than rectangular ones but take longer to print. Rectangular infills work better for high infills.

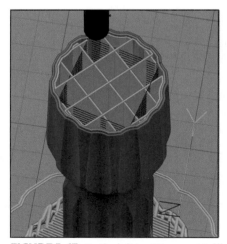 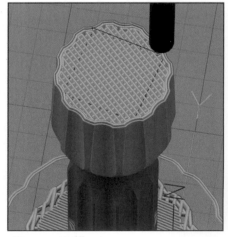

FIGURE 5-17: On the left is a 10 percent infill; on the right is a 50 percent infill.

REDUCE PRINT TIME

The longer a file takes to print, the greater the chance of something going wrong. When printing multiple parts at once, place them between 5 mm and 15 mm apart so that the extruder doesn't have as far to travel between them. This reduces build time. That said, placing parts too close together is risky because if one comes loose it can get dragged by the extruder and ruin the others. It's best to limit the number of parts printed at one time.

Splitting a large part in half is also prudent for models that take a long time to print. The smaller parts will take less time to print, lessening the risk of having a large print fail after many hours have been spent printing it. Minimizing supports also reduces print time and saves material.

ANALYSIS SOFTWARE

Visually inspect all files when you're done modeling them. Setting the display view to wireframe mode or applying a glass (transparency) appearance can help with this. Afterward, run analysis software on the file because it can find problems not apparent with a visual inspection, such as tiny holes, overlapping faces and vertices, or reversed faces. This can save you a lot of time and plastic wasted trying to print defective files.

FIGURE 5-18: The Meshmixer Analysis options

There are many analysis programs available (see Additional Resources). Of particular note is Autodesk Meshmixer, a free, stand-alone program at **meshmixer.com**. It has Analysis icons that check for defects, strength optimal orientation, and other features (Figure 5-18).

Sometimes one analysis program can fix all problems. Sometimes it just identifies problems, requiring you to return to Fusion to fix or try to fix in another program. Figure 5-19 shows a vase inside Meshmixer that had a lot of problems. Meshmixer fixed all but one and identified it with a red pin. At this point we could export it from Meshmixer as an STL file and either reimport into Fusion to fix or import into another analysis program that might be able to fix it.

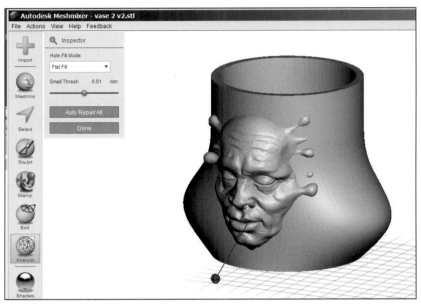

FIGURE 5-19: Meshmixer fixed all but one defect in this file.

Netfabb for Fusion

Netfabb is an analysis program that has free and pay versions. The Netfabb for Fusion version is free. To install it, go to the Model workspace. Click Add-Ins ➡ Scripts And Add-Ins (Figure 5-20), and on the Add-Ins tab, click NetfabbForFusion (Figure 5-21).

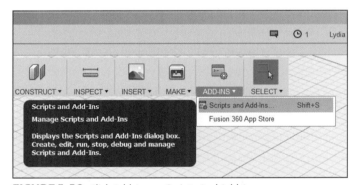

FIGURE 5-20: Click Add-Ins ➡ Scripts And Add-Ins.

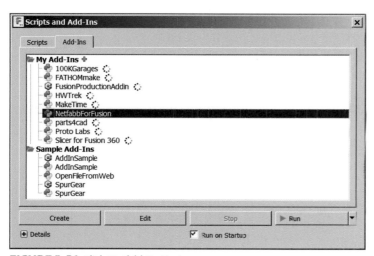

FIGURE 5-21: Click NetfabbForFusion.

If NetfabbforFusion isn't listed, click Add-Ins → Fusion 360 App Store (Figure 5-22). This takes you to the app store website (Figure 5-23). Search for Netfabb (Figure 5-24), and then download it and run the file. It won't appear as an icon on your desktop upon completion, but you should see it listed under Add-Ins → Scripts And Add-Ins now.

FIGURE 5-22: Click Add-Ins → Fusion 360 App Store.

FIGURE 5-23: The Autodesk app store website

FIGURE 5-24: Search for Netfabb.

Click the Run button at the bottom of the Add-Ins window. A dialog box with an installation link will appear (Figure 5-25). Install it. Then select the model you want to send to Netfabb and click Make ➜ Netfabb for Fusion 360 (Figure 5-26). It will send the model to that program, converting any solid bodies to meshes in the process. You can choose a tessellation level from Fine (the default) to Coarse.

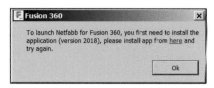

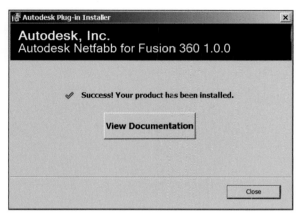

FIGURE 5-25: A dialog box with an installation link will appear when you click the Run button.

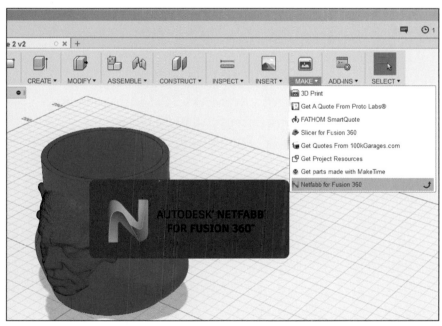

FIGURE 5-26: Select the model and click Make → Netfabb For Fusion 360.

Find tutorials for using Netfabb at the Autodesk Netfabb YouTube channel. And now that you've learned how to use Fusion 360's basic tools and where to find some analysis programs, let's make stuff.

Additional Resources

Message boards for 3D printing talk: 3D Hubs Talk

Facebook groups: 3D Printing, 3D Printing Club

Stratasys best practices suggestions: **www.stratasys.com/resources/best-practices**

STL Analysis Software:

Netfabb free and pay options: **www.netfabb.com**

Simplify3D, a for-pay slicer that contains analysis functions: **www.simplify3d.com**

Tinkercad, a free modeling web app that fixes STL files upon import: **www.tinkercad.com**

Meshmixer, a free modeling program that has analysis and repair functions: **www.meshmixer.com**

Autodesk Print Studio. This is a legacy program that is no longer supported, but it's a good program. A Windows version can still be downloaded at **https://support.ember.autodesk.com/hc/en-us/articles/212823998-Install-Print-Studio**.

Part II

MAKE SOME STUFF

EMOJI WALL ART

In this chapter, we'll turn a web image into wall art using Fusion's Model workspace, a digital imaging program, and an online converter.

FIND AN IMAGE

I searched online for emojis and found the JPEG file in Figure 6-1. Simple cartoons with contrasting colors convert to SVG files best. Black-and-white ones convert better than color ones, as details are usually better preserved. So I imported the file into IrfanView (download at **irfanview.com**), clicked Image → Convert To Grayscale (Figure 6-2), and saved it as a PNG file. PNG files preserve transparencies, resulting in a clear background.

FIGURE 6-1: A web image

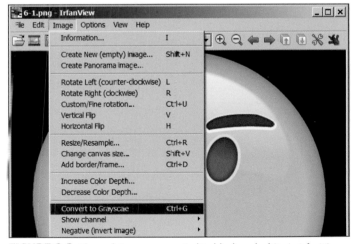

FIGURE 6-2: The web image converted to black and white in IrfanView

USE AN ONLINE CONVERTER TO MAKE AN SVG FILE

Point your browser to **online-convert.com**. Click the dropdown arrow in the Image Converter box, choose Convert To SVG, upload the file, and then save it as an SVG file (Figure 6-3).

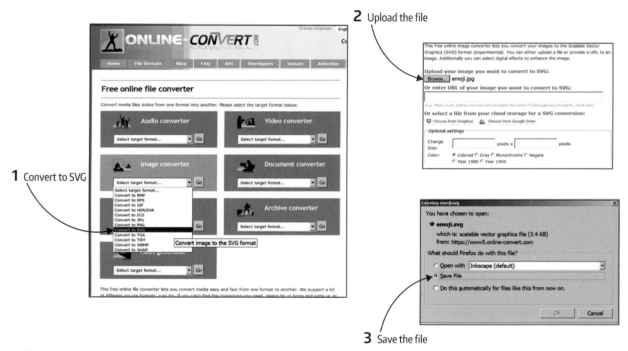

FIGURE 6-3: Convert the PNG file to an SVG file.

IMPORT AND UNLOCK THE SVG FILE

FIGURE 6-4: Click Insert → Insert SVG.

Click Insert → Insert SVG (Figure 6-4). The origin planes and a dialog box will appear. Click the horizontal plane and then click the folder graphic. Navigate to the SVG file, bring it in, and click OK. The file will be green, indicating that it's locked (Figure 6-5). Select it, right-click, and choose Fix/UnFix (Figure 6-6). The sketch will turn blue, indicating it is now unlocked and editable. Be aware that depending on the graphics program you use, or the conversion you get, your SVG might look different from mine and require a bit different editing.

Folder graphic

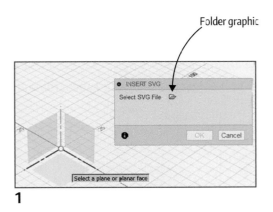

1

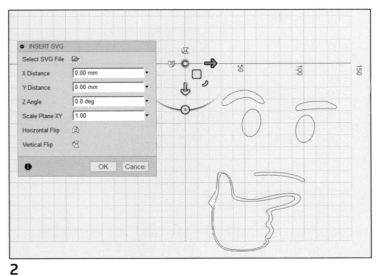

2

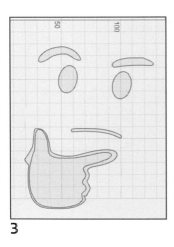

3

FIGURE 6-5: Import the SVG file.

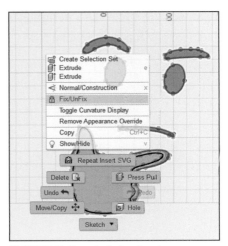

FIGURE 6-6: Unlock the sketch.

ADD A CIRCLE

The sketch needs a circle around it. Click Circle → Center Diameter Circle, click a center point, and then click or type a diameter (Figure 6-7).

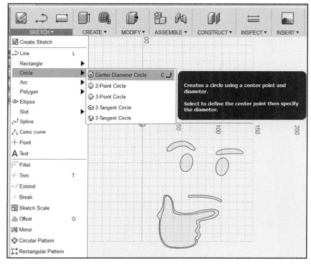 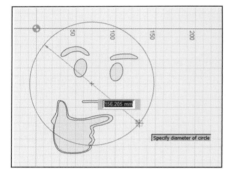

FIGURE 6-7: Draw a circle around the sketch.

MODEL THE SKETCH

Select the hand and facial features by pressing and holding the Shift key and clicking each. Then right-click, choose Press Pull, and extrude those sketches up (Figure 6-8). The circle sketch will disappear; click the gray light bulb in front of its entry in the Browser to make it yellow, turning it back on (Figure 6-9). Select the circle sketch, right-click, and press-pull it up half the distance of the hand and facial features (Figure 6-10).

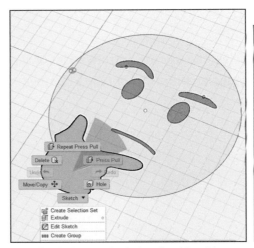

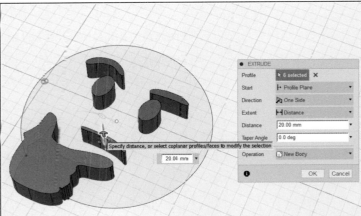

FIGURE 6-8: Extrude the hand and facial features up.

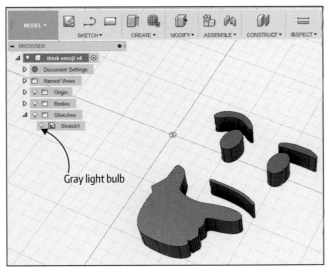

Gray light bulb

FIGURE 6-9: Turn the circle sketch back on by clicking the gray light bulb in front of its Browser entry.

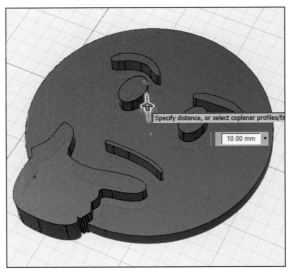

FIGURE 6-10: Press-pull the circle up half the distance of the facial features.

FILLET THE MODEL

Press and hold the Shift key and select the top and bottom edges of the circle. Right-click, choose Fillet, and drag the arrow or type a number for the desired curve (Figure 6-11).

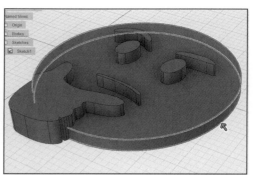 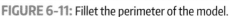 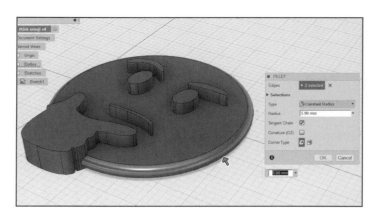

FIGURE 6-11: Fillet the perimeter of the model.

USE THE TIMELINE TO ADD A NAIL HOLE

Oops! We forgot to make a nail hole on the back of the plaque. Use the timeline to do this. Roll the slider to the left until just the emoji sketch appears (Figure 6-12). You might have to click the light bulb for the sketch to appear. Then click the Circle tool (or the Slot tool for a narrower hole), click the sketch to select it, and draw a circle on it (Figure 6-13). After placing the circle you can still move it later, if needed, to finesse its position.

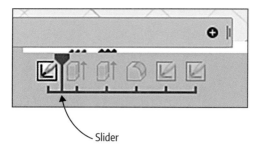

Slider

FIGURE 6-12: Roll the timeline slider back until just the sketch appears.

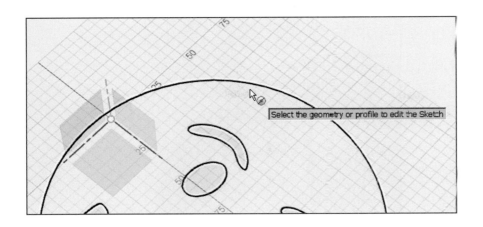

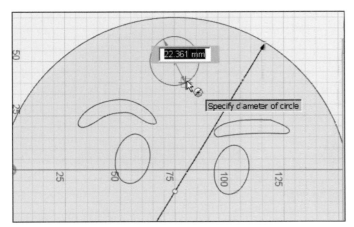

FIGURE 6-13: Select the sketch and then draw a circle on it.

Now roll the timeline slider to the right until the body appears again (Figure 6-14). Select the circle (you might have to hide the body or rotate it to do this), right-click, choose Press Pull, and push it a little into the back (Figure 6-15). Done! Well, almost. Right-click the root and choose Save As STL (Figure 6-16). Figure 6-17 shows the printed file.

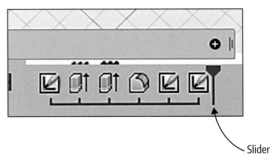

Slider

FIGURE 6-14: Roll the timeline slider forward until the body appears again.

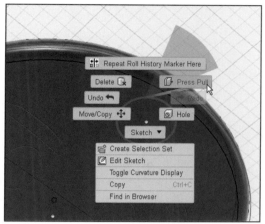

1

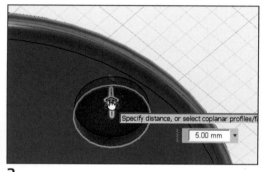

2

3

FIGURE 6-15: Push the circle into the back of the body.

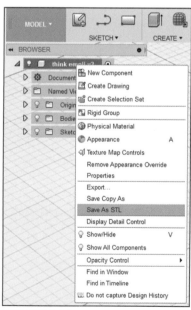

FIGURE 6-16: Choosing Save As STL

FIGURE 6-17: The printed file

TWO CHAIRS AND A VASE

In this chapter, we'll work in the Sculpt workspace and take a couple of short detours through the Patch and Mesh workspaces to make some organic-shaped chairs and a vase.

TALL-BACK CHAIR

Click the Create Form icon to enter the Sculpt workspace and then choose Create → Box (Figure 7-1). Make a box equally sized on all axes (Figure 7-2).

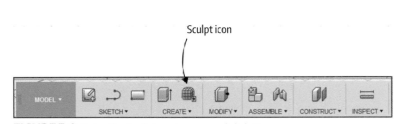

FIGURE 7-1: Enter the Sculpt workspace and click Box.

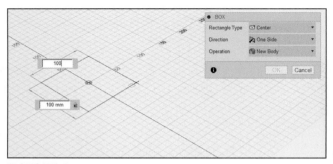 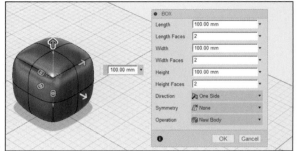

FIGURE 7-2: Make a box of equal proportions.

Edit Form

Select two faces on the top, as shown in Figure 7-3. Right-click, and then choose Subdivide. The two faces will turn into eight (Figure 7-4).

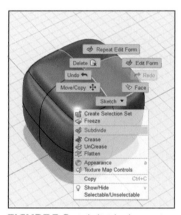 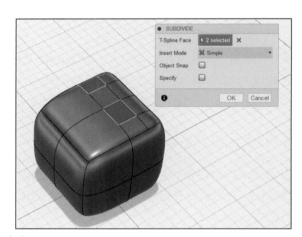

FIGURE 7-3: Subdivide the top into multiple faces.

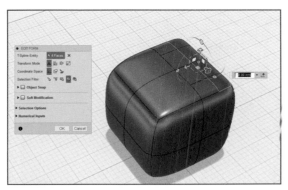

FIGURE 7-4: Select faces, choose Edit Form, and pull the faces up.

Select the edge faces, right-click, and choose Edit Form. A widget will appear; pull the selected faces up. Then select the seat, right-click, and choose Edit Form. Push the face down (Figure 7-5).

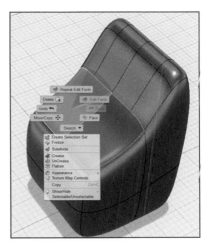
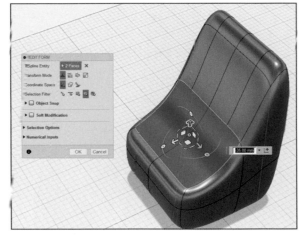

FIGURE 7-5: Use Edit Form to push the chair's seat down.

Select and right-click the seat edge shown in Figure 7-6 and choose Insert Edge. Drag the handle to place the new edge below the original one and click OK. The chair's shape will change a bit (Figure 7-7).

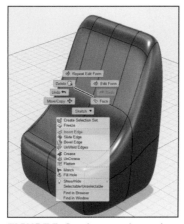

FIGURE 7-6: Select an edge, right-click, and choose Insert Edge.

FIGURE 7-7: Change the chair's shape with an inserted edge.

Bevel and Crease

Finesse the chair's shape. Select an edge on the back and bevel it (Figure 7-8).

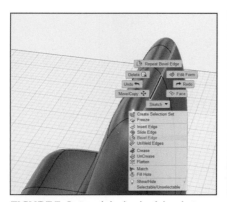

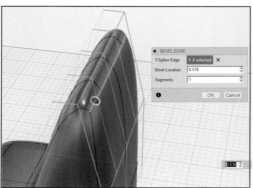

FIGURE 7-8: Bevel the back of the chair.

Then select an edge where the seat and back meet, right-click, and choose Crease (Figure 7-9). That will put a hard edge there instead of a rounded one (Figure 7-10).

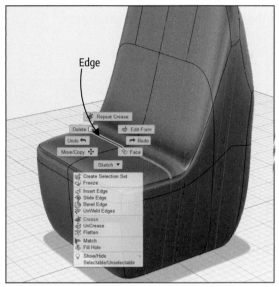

FIGURE 7-9: Select an edge and choose Crease.

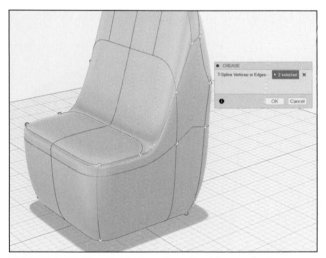

FIGURE 7-10: The chair's shape has a hard (creased) edge now.

SHORT-BACK CHAIR

Enter the Sculpt workspace and choose Create → Box, and then make a box equally sized on all axes (Figure 7-11). Then enter the Patch workspace (Figure 7-12). The subdivisions on the box will disappear.

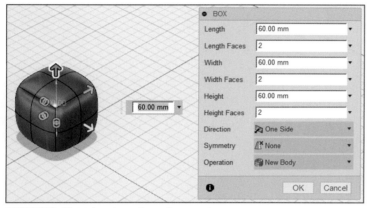

FIGURE 7-11: Make a box of equal proportions.

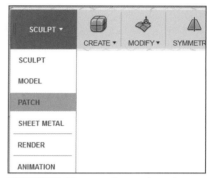

FIGURE 7-12: Enter the Patch workspace.

Delete Faces

Select the two adjacent faces shown in Figure 7-13 and press Delete.

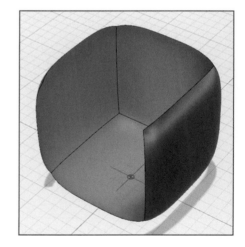

FIGURE 7-13: Delete the two adjacent faces.

Thicken Faces and Round the Edges

Select any remaining face, right-click, and choose Thicken. Then drag the arrow or type a specific thickness (Figure 7-14). To round the edges, press and hold the Shift key and select the ones shown in Figure 7-15. Right-click, choose Fillet, and type or drag a fillet radius.

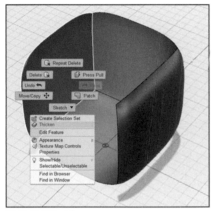

FIGURE 7-14: Thicken the chair.

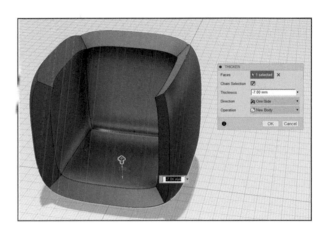

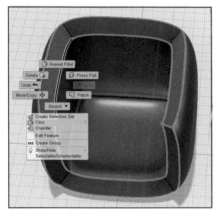

FIGURE 7-15: Select and fillet the edges.

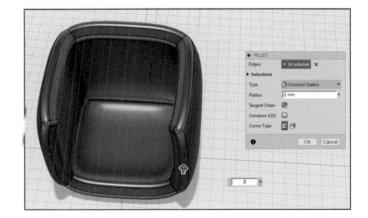

Change Proportions

Let's change this chair's proportions. Drag a selection window around it. Click Modify ➜ Scale, and set the dropdown arrow in the dialog box to Non Uniform (Figure 7-16). Then push the z-axis arrow down (Figure 7-17).

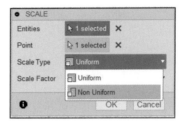

FIGURE 7-16: Click Scale ➜ Non Uniform.

FIGURE 7-17: Adjust the height with the z-axis arrow.

VASE

Enter the Sculpt workspace and choose Create→ Cylinder; then make a tall cylinder. Double-click an edge at the bottom to select the whole ring (Figure 7-18). Right-click one of the selected edges, choose Edit Form, and drag the ring out with the manipulator's center button to deform it (Figure 7-19).

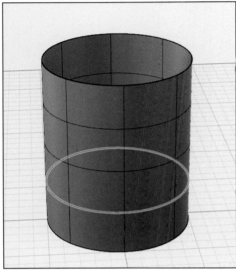

FIGURE 7-18: Make a cylinder and double-click an edge to select the whole ring.

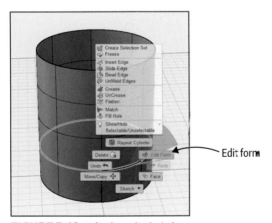

FIGURE 7-19: Edit the cylinder's form.

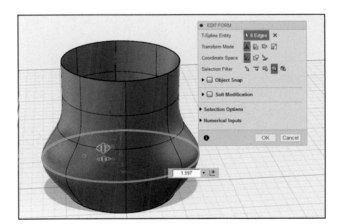

Fill and Straighten the Bottom

Double-click an edge on the bottom of the cylinder to select the whole ring. Then click Modify → Fill Hole (Figure 7-20). The hole will close. The bottom will be rounded, so click Modify → Crease to make it flat (Figure 7-21).

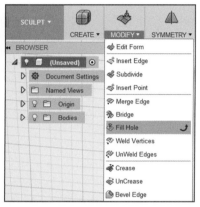
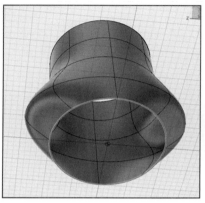
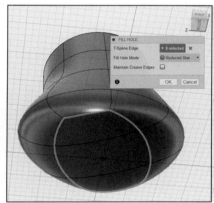

FIGURE 7-20: Fill the hole at the bottom of the cylinder.

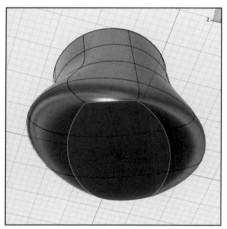
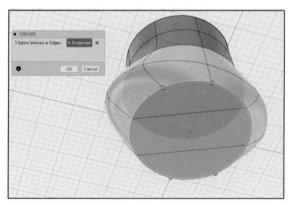

FIGURE 7-21: Crease the bottom to make it flat.

Attach an STL File to the Vase

Let's jazz this vase up with an imported STL file of a face. Figure 7-22 shows a file I downloaded from **thingiverse.com**. Enter the Model workspace and click Insert → Insert Mesh (Figure 7-23). Use the manipulators to position it as shown in Figure 7-24. Scale it down if needed by selecting it and clicking Modify → Scale.

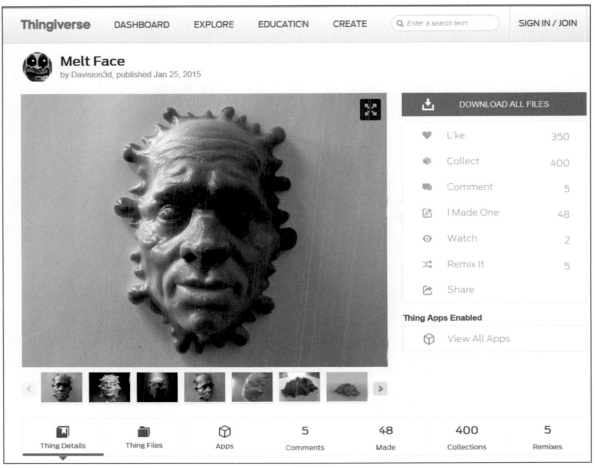

FIGURE 7-22: A file at thingiverse.com

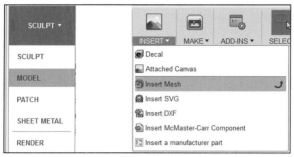

FIGURE 7-23: Insert the STL file.

FIGURE 7-24: Position the STL file on the vase.

Reduce the STL File's Polygon Count

The face file has a very large number of polygons, as you can see from
the dense mesh. This means the Mesh to BRep conversion tool won't
work, so the file can't be edited. To make the polygon count smaller,
enter the Mesh workspace, if Fusion didn't already default to it when

you inserted the file (Figure 7-25). If the Mesh workspace doesn't appear, you might have to enable it in the Preferences menu.

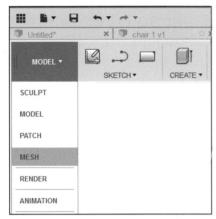

FIGURE 7-25: Enter the Mesh workspace.

Select the face through its Browser entry and then click Modify → Reduce. A dialog box will appear (Figure 7-26).

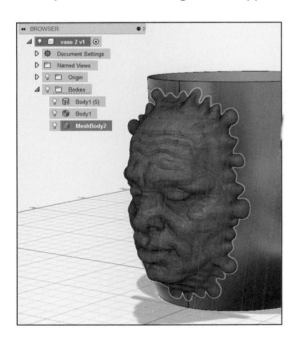

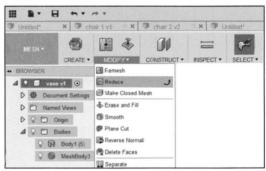

FIGURE 7-26: Select the file and click Modify → Reduce.

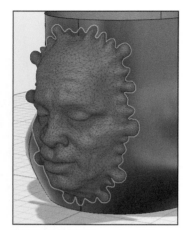

The dropdown menus offer adaptive (preserves the shape) or uniform (removes polygons uniformly) options. You can also adjust density, face count, and tolerance—all three, not just one. I changed the density, which controls how close the changed polygon edges are to the original edges. The smaller the number you enter, the less dense the file will be. I typed **.10**; Figure 7-27 shows the result.

Optional: Convert to BRep

You can edit the file in the Mesh menu or you can convert it to a solid file to edit it in the Model workspace. It just depends on the kind of edits you want to do. To alter it as a solid, select it, right-click, and choose Mesh to BRep. If the file has a low enough polygon count and has no flaws, it will convert (Figure 7-28). You might get a message saying it will take a while; click and in a few minutes you should have your result.

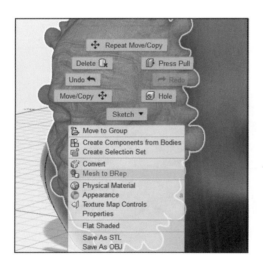

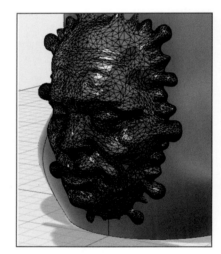

FIGURE 7-28: Convert the mesh file to a solid file if you want to edit it in the Model workspace.

Thicken in Model Space

Thicken the vase now. Thickening it in the Model (parametric) work-space enables you to edit it later. Select both the bottom and one side face (press and hold Shift), right-click, and choose Thicken. Then drag the arrow or type a number (Figure 7-29). In this case, I thickened it enough to cover the portion of the face that protruded through the vase's interior.

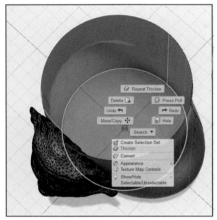
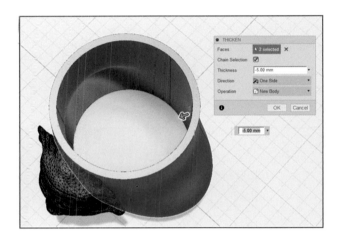

FIGURE 7-29: Thicken the vase.

Repair Flaws

Whether the face file is kept as a mesh or converted to solid, it, like most meshes, typically needs some repair to be 3D-printable. One program might fix it, or you might have to fix it with multiple programs.

In this case, I converted the face file to a solid with the Mesh to BRep tool. Then I exported the whole vase as an STL file and imported that STL file into Autodesk Meshmixer (**meshmixer.com**). There, I ran the Inspector tool. It found multiple flaws and fixed all but one (Figure 7-30). I exported it as an STL file from Meshmixer and imported it into Tinkercad (**tinkercad.com**), an Autodesk web app

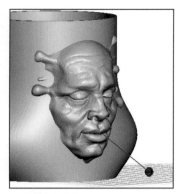

FIGURE 7-30: The red pin shows the flaw Meshmixer couldn't fix.

that automatically fixes STL files upon import (Figure 7-31). Then I exported an STL file from Tinkercad and reimported it into Meshmixer to make sure it was fixed. That flaw was indeed fixed, but Tinkercad added a few more flaws! Meshmixer was able to fix those, however, and I exported it as an STL. Finally, I sliced that STL with MakerBot software and printed it on a MakerBot Replicator 2 with Hatchbox filament (Figure 7-32).

> **TIP** If both the vase and face don't export to an STL file, convert the face mesh to a solid and Combine → Join the vase and face together.

FIGURE 7-31: The file imported into Tinkercad, which fixes flaws upon import.

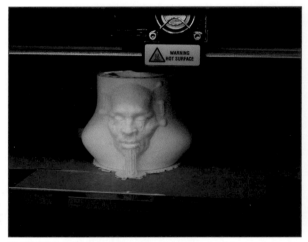

FIGURE 7-32: The vase under construction with supports, raft, and a 10% infill

8

FIDGET SPINNER

In this chapter, we'll make a fidget spinner using a #608RS skate ball bearing as the spinner and three quarters as the weights (Figure 8-1).

FIGURE 8-1: A ball bearing and three quarters

SKETCH A CIRCLE FOR THE BALL BEARING

The ball bearing's diameter is 22 mm, so sketch a circle on the horizontal plane just a bit bigger than that, 22.1 mm (Figure 8-2).

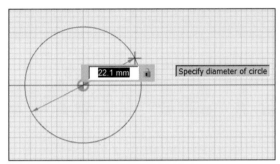

FIGURE 8-2: Sketch a circle a little bit bigger than the ball bearing's diameter.

SKETCH A CIRCLE FOR THE QUARTER

Hover your mouse over the center of the bearing's circle and move the mouse up to get a vertical inference line. This will enable you to align the center of the second circle with the first (Figure 8-3). Alternatively, you can align two circles by clicking the Horizontal/Vertical constraint onto the circles' centers (Figure 8-4). The initial distance between the circles doesn't matter because you will adjust it later.

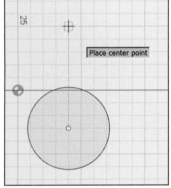

FIGURE 8-3: Align the circles' centers by hovering over the first circle's center to get an inference line.

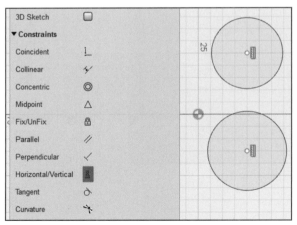

FIGURE 8-4: Click the Horizontal/Vertical constraint onto both circles' centers to align them.

Once you click the center point, set the diameter. A quarter is 24.26 mm in diameter, so make the circle a bit bigger, 24.36 mm.

SPACE THE CIRCLES 30 MM APART

Choose Sketch → Sketch Dimension. Then click the circles' centers and drag a dimension line off to the side. Type **30** in the text field and press Enter. The dimension line will adjust to 30 mm between the centers (Figure 8-5).

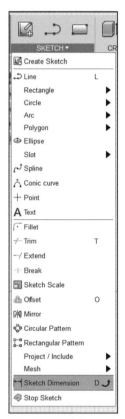

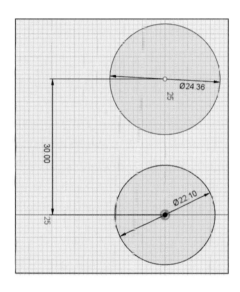

FIGURE 8-5: Space the circles' centers 30 mm apart.

OFFSET THE CIRCLES

Offset both circles 3 mm by clicking Sketch ➔ Offset and either dragging the handle or typing **3mm** in the text field (Figure 8-6).

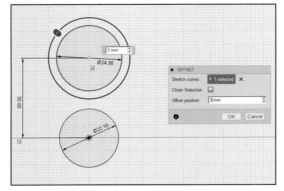 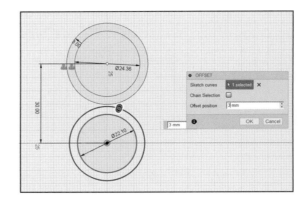

FIGURE 8-6: Offset both circles 3 mm.

COPY AND ARRAY A CIRCLE

Copy the quarter circle three times around the bearing circle. Click Sketch ➔ Circular Pattern, and select both the inner and outer circles (you don't need to hold Shift to select them both). Click the Center Point button in the dialog box, and then click the bearing circle's center. Type **3** in the quantity box (Figure 8-7).

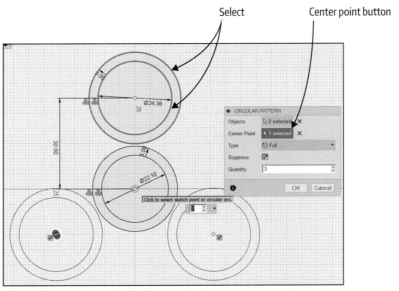

FIGURE 8-7: Use the Circular Pattern tool to copy and array the quarter circles around the bearing circle.

SKETCH A 3-POINT ARC BETWEEN THE CIRCLES

Choose Sketch ➜ Arc ➜ 3-Point Arc and sketch one as shown in Figure 8-8 by clicking it on the quarter circles and then clicking it onto the bearing circle. Make sure the tangent constraint appears on the bearing circle; if it doesn't, apply it from the Constraints palette.

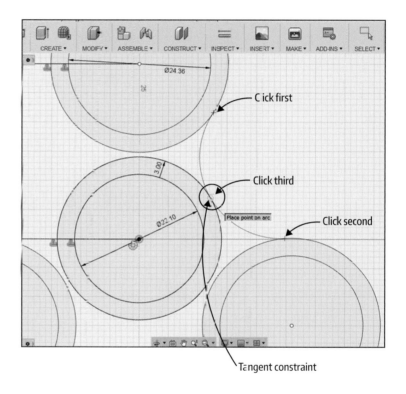

FIGURE 8-8: Draw a three-point arc.

COPY AND ARRAY THE ARC

Copy the arc three times around the center of the bearing circle with the Circular Pattern tool (Figure 8-9). To do this, select the arc, click the tool, click the Center Point button in the dialog box, and then click in the center of the bearing circle.

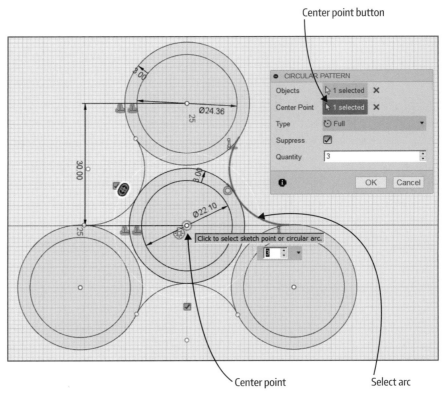

FIGURE 8-9: Use the Circular Pattern tool to copy the arc around the bearing circle.

EXTRUDE THE SKETCH

Click Stop Sketch and the model will return to a 3D view. Hold and press Shift and select the geometry shown in Figure 8-10. Right-click a selected part and choose Press Pull. Extrude the whole sketch up 7 mm, the height of the bearing.

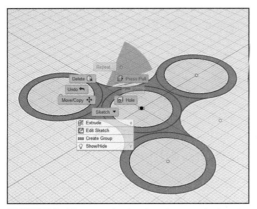 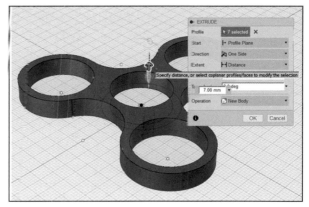

FIGURE 8-10: Extrude the sketch up to the height of the bearing.

ROUND OFF THE EDGES

Choose Modify → Fillet. Then drag a selection window around the whole model (Figure 8-11). Drag the arrow or enter **1 mm** for the fillet radius in the text field, and click OK to finish (Figure 8-12). Done!

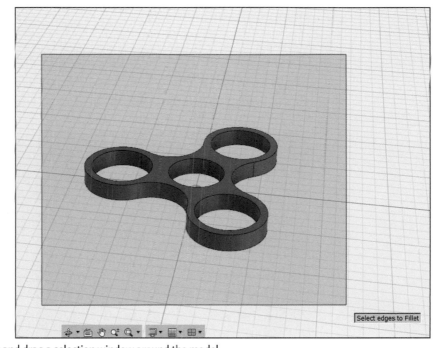

FIGURE 8-11: Click Fillet and drag a selection window around the model.

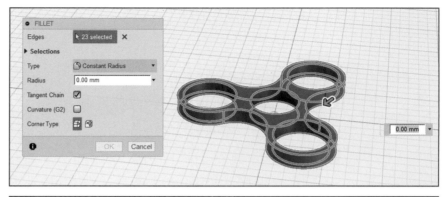

FIGURE 8-12: Enter a fillet radius.

I printed it on a MakerBot Mini with MakerBot filament. It was a bit too small to fit the coins and bearing, so I scaled the whole thing up 2 mm in the MakerBot slicer. That worked (Figure 8-13).

FIGURE 8-13: The printed fidget spinner

9

GEARS

In this chapter, we'll do two things. First, we'll download and edit a gear from the McMaster-Carr catalog. Second, we'll install and use Fusion's gear generator and make a planetary gear assembly with it.

DOWNLOAD A GEAR FROM THE MCMASTER-CARR CATALOG

McMaster-Carr is an industrial products supplier. Many of the things it sells are in CAD format, and a library is linked to Fusion. To browse it, choose Insert → Insert McMaster-Carr Component (Figure 9-1). A window will appear; I searched for gears (Figure 9-2) and scrolled through the hits. You can also search by specific features, such as gear pressure angle, pitch, and number of teeth.

FIGURE 9-1: Click Insert → Insert McMaster-Carr Component.

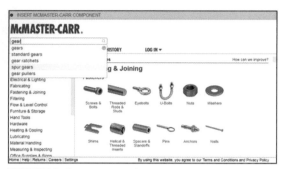

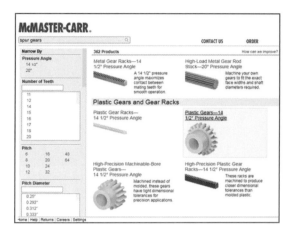

FIGURE 9-2: A gear search

Select a product, and if it has downloadable files (not all products do), you'll see a green CAD graphic in the Product Detail box (Figure 9-3). Click the graphic and scroll down to find the text field that lists downloadable files (Figure 9-4). This text field is typically on top or to the right of the product drawing. Once you find it, click the drop-down arrow for file format choices. Then click Save to import it into Fusion.

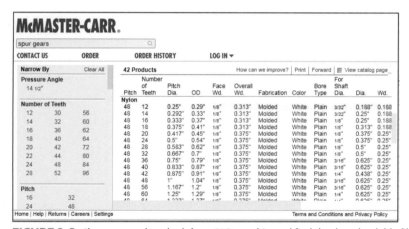

CAD graphic

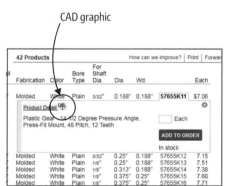

FIGURE 9-3: Choose a product, look for a CAD graphic, and find the downloadable files.

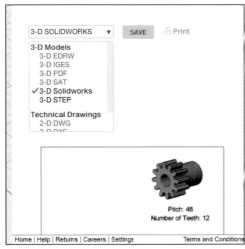

FIGURE 9-4: Choose the file format and click Save.

EDIT THE MCMASTER-CARR GEAR

Figure 9-5 shows the gear I downloaded. It has an attached base. The base can be deleted, but any fillet or chamfer features on it must be deleted first. After deleting those features, select the base and delete it (Figure 9-6). Note that the hole remains. You can delete the hole's fillet and then the hole itself before deleting the base, which will result in the gear's hole being deleted, too (Figure 9-7).

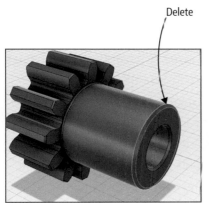
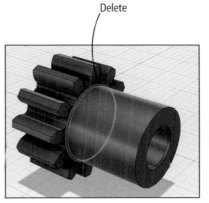
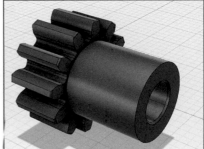

FIGURE 9-5: Delete features on the base.

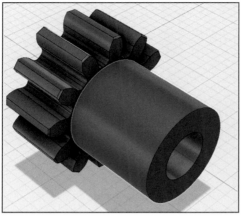

FIGURE 9-6: Delete the base after deleting its features.

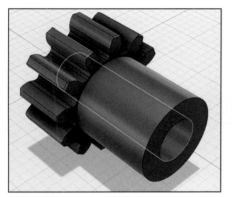

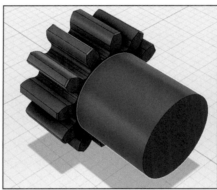

FIGURE 9-7: Delete the hole to remove it from the gear.

You can further edit the gear—for example, make it thinner by extruding its face (Figure 9-8). You can also sketch and extrude circles on top of it to make holes. If you want to access this gear from the timeline, its features are grouped. This means that all the individual operations are contained inside one icon. Expand the group by right-clicking the gear's timeline icon and choosing Expand Group (Figure 9-9). You can also just click the plus sign under the icon in the timeline to expand the group. When using a lot of repeating products, such as screws, suppress the threads (right-click the timeline icon) to make Fusion run faster. Deleting them adds a lot of icons to the timeline.

Drag a window around the McMaster-Carr gear and delete it now, since we don't need it anymore.

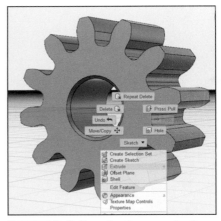
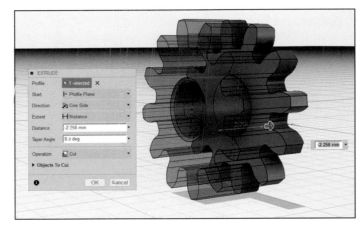

FIGURE 9-8: Extrude a face to make the gear thinner.

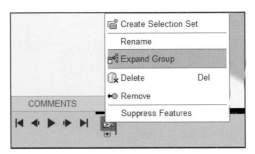
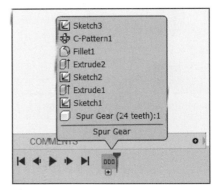

FIGURE 9-9: Right-click the gear's timeline icon to expand its features.

GEAR GENERATORS

Designing gears involves calculating module, number of teeth, pressure angle, and more. This is beyond the scope of this book; some references that explain how to do this are at the end of this chapter. Once you have this data, you can either draw the gear manually or use a gear generator. There are websites that will generate gears with your data. Autodesk's App Store has gear generators made by third-party developers (Figure 9-10) that you can purchase and install. Access the app store under the Add-Ins menu or directly at **apps.autodesk.com/en**. There's also a simple gear generator inside Fusion, which we're going to use.

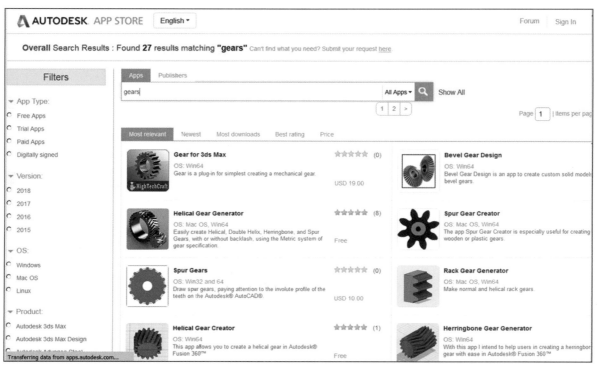

FIGURE 9-10: Gear generators in the Autodesk App Store

ADD FUSION'S SPUR GEAR TOOL

Choose Add-Ins → Scripts And Add-Ins, and then select the Add-Ins tab in the resulting dialog box (Figure 9-11). Click the second Spur Gear entry and click Run. This puts a Spur Gear tool in the Create menu of the Model workspace (Figure 9-11), which is good until you exit Fusion. The first Spur Gear entry simply runs the add-in and stops when you're finished.

> **TIP** What's the difference between a script and add-in? A script runs and then it's done. You execute it through the Scripts And Add-Ins command. It stops immediately after the Run function is executed. An add-in is automatically loaded when Fusion starts up.

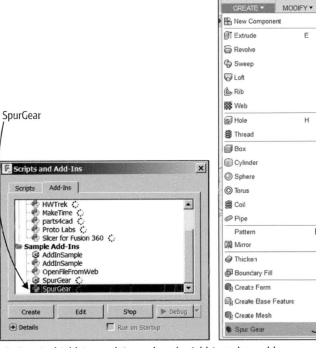

SpurGear

FIGURE 9-11: Click Add-Ins → Scripts And Add-Ins, and then select the Add-Ins tab to add a spur gear to the Create menu.

MODEL A PLANETARY GEAR TRAIN

Also called an epicyclic gear train, a planetary train has a ring, a sun, and planets (Figure 9-12). The sun is fixed and the planets revolve around it. The planets mesh with the sun and ring. Watches and pencil sharpeners are common items that use planetary gears. Our gear train will have a sun gear, three planet gears, and a ring.

Ring

Sun

Planet

FIGURE 9-12: Planetary gears

Generate the Spur Gears

Click the Spur Gear tool to open its dialog box, which contains default parameters that you can replace with data you calculate yourself. Our project needs five gears: three planets, one sun, and one ring. They'll all have the same data except for number of teeth. The ring gear has 96 teeth, the sun gear has 24 teeth, and the planets each have 35 teeth. Figure 9-13 shows the screen with the ring gear's parameters.

> **TIP** The same set of gear parameters may yield different results in different generators. For example, gears that mesh well on a website generator or in another software program may need to have some parameters, such as Module and Fillet Radius, finessed to mesh correctly when generated in Fusion.

SPUR GEAR dialog

Module: Size Ratio
(Pitch Diameter /
Number of Teeth)

Pitch Diameter

Hole Dia.

Root fillet radius

Standard	Metric ▾
Pressure Angle	20 deg ▾
Module	1.00 ▾
Number of Teeth	96
Backlash	0.00 mm ▾
Root Fillet Radius	0.00 mm ▾
Gear Thickness	2.70 mm ▾
Hole Diameter	1.00 mm ▾
Pitch Diameter	96.00 mm

OK   Cancel

FIGURE 9-13: The ring gear parameters

The gears enter the workspace at the same location. Select them through their Browser entry, right-click, choose Move → Copy, and move them off the one underneath them (Figure 9-14). Figure 9-15 shows the gears I generated. All have a green dashed circle around them, which is called the pitch diameter circle. That's where another circle meshes with it; this is useful if you draw gears and teeth manually.

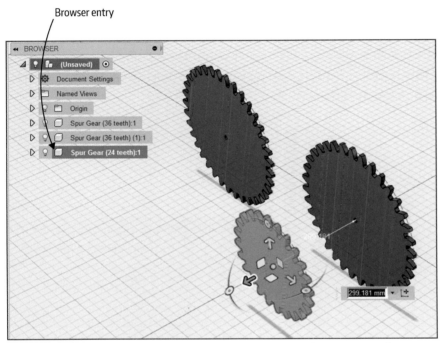

FIGURE 9-14: Move the gears off each other. The sun and two planet gears are shown.

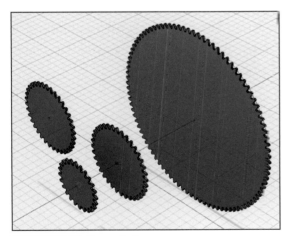

FIGURE 9-15: Generated gears

Model a Ring Gear

If the workspace isn't already on Orthographic mode, set it to that mode to make sketching easier (Figure 9-16), and view the model from the top. Be aware that as you work you'll probably get an occasional message: "Some components have been moved." Just click the Capture Position button when that happens (Figure 9-17). This message simply means that you moved the component; clicking the Continue button records the moved distance in a Snapshot operation that goes into the timeline. Clicking Revert And Continue undoes the move. This is probably most useful when making animations or versions of models in different positions.

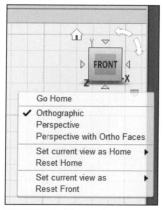

FIGURE 9-16: Set the workspace to Orthographic mode.

FIGURE 9-17: Click the Capture Position button when this message appears.

Choose Sketch → Circle → Center Diameter Circle. Click first on the big gear we're using for the ring, second on the gear's center, and third on a place off the gear to set the diameter (Figure 9-18). The diameter should be bigger than the gear's diameter.

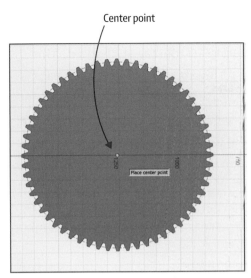

Center point

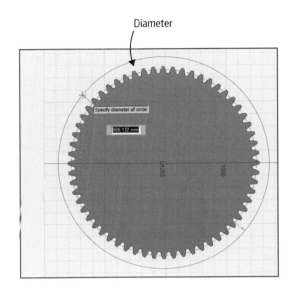

Diameter

FIGURE 9-18: Sketch a circle on the ring gear.

Select and extrude the circle forward around the gear. Then select the gear through the Browser, right-click, and click Delete (Figure 9-19). You'll get a warning message that the feature is referenced by other features, but you should be able to continue. The gear will be deleted, and you'll be left with a ring that has 96 teeth.

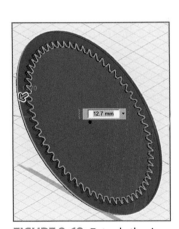

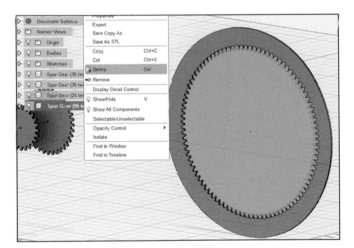

FIGURE 9-19: Extrude the ring and delete the gear.

Position the Gears

Now move the gears into a planetary arrangement. If you do this manually, deselect the Incremental Move box in the Grid And Snaps settings at the bottom of the workspace (Figure 9-20) to make moving them easier. You can also use the Create menu's Circular Pattern tool to position the planet gears around the sun gear.

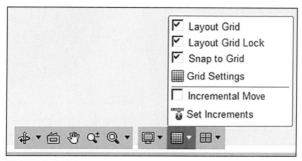

FIGURE 9-20: Deselect Incremental Move to make manually positioning the gears easier.

Turn the gear ring into a component by right-clicking its Body entry in the Browser and choosing Create Components From Bodies. To make an assembly, right-click the ring component and choose Ground. Then add As-Built revolute joints (Figure 9-21) on each of the other gears, using any circle on them as the axis to place the joint on (Figure 9-22).

Done! When 3D printing this, print all the gears separately so that they don't fuse together during the printing process.

> **TIP** McMaster-Carr components can be used to make stamps (impressed patterns).

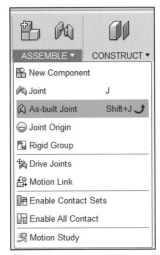
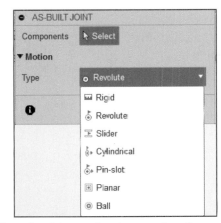

FIGURE 9-21: Make an assembly with revolute joints to place the gears together.

FIGURE 9-22: The planetary gear assembly. A revolute joint is shown on the sun gear.

Additional Resources

Gear generators: **http://hessmer.org/gears/InvoluteSpurGearBuilder.html**,

https://woodgears.ca/gear_cutting/template.html

How to Determine Gear Ratio:
www.wikihow.com/Determine-Gear-Ratio

10

CAR PHONE MOUNT ASSEMBLY

In this chapter, we'll work off a pencil sketch to make a dashboard mount for a cellphone. Despite the project's complexity, this will be a loose approach because it emphasizes working through a prototype idea in your head. This is useful for the conceptual phase of product development. Our intent is to generate a concept model, and for this purpose it doesn't have to be dimensionally perfect. So you will use your own dimensions in this project—you can be the engineer doing trial and error.

Google the dimensions of your phone and hand-sketch an idea for a mount. Proportional accuracy of the sketch is important, so use a ruler or grid paper. Figure 10-1 shows orthographic drawings for an iPhone 6 mount. It consists of a stand and clip, held together with a socket and ball joint. I added 2 mm to the phone's height and width dimensions to allow space for the phone to fit inside the clip. You can model this by typing all needed dimensions as you work, or you can trace the sketch and scale it at the end. We'll do the latter.

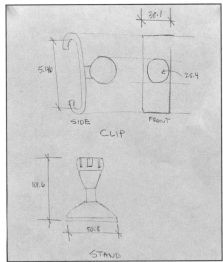

FIGURE 10-1: Sketches of a dashboard mount for an iPhone 6

IMPORT THE HAND SKETCH

To bring the sketch into the workspace, click Insert ➜ Attached Canvas. (If you want to use my sketch, take a photo of it and use that as the file.) Select a face to attach the sketch canvas to. Then click the picture graphic to browse to the file. Press Enter to bring it in. Arrows will appear that you can use to move the file, but in this case keep it where it is—at the origin. Click OK to finish (Figure 10-2).

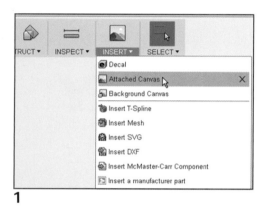

1

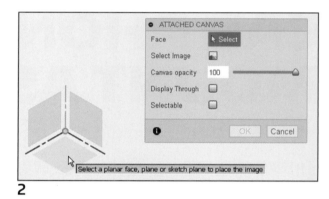

2

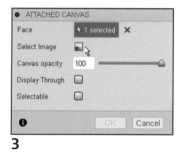

3

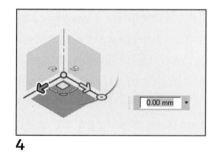

4

FIGURE 10-2: Insert the hand sketch into the workspace.

You can only place the sketch in the x-y plane. If you want to place it above the x-y plane, offset a plane from the origin's x-y plane and attach the sketch to it when inserting (Figure 10-3). You can also attach the sketch to an existing body instead of on the work plane. If you need to move the sketch later, right-click its Browser entry to select it and choose Edit Canvas (Figure 10-4). Then right-click the canvas and choose Move.

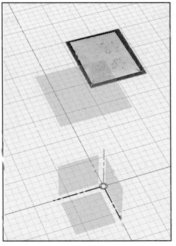

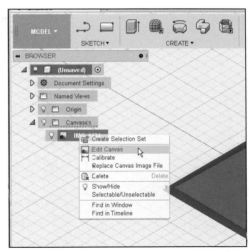

FIGURE 10-3: To insert the canvas above the x-y plane, offset a plane and attach the canvas to it.

FIGURE 10-4: To move the canvas after insertion, right-click its Browser entry and click Edit Canvas.

Note that one of the Edit Canvas options is Opacity (Figure 10-5). If the sketches you draw on top of it "disappear" or you want to trace an imported canvas, move the opacity slider left to make it transparent.

Opacity

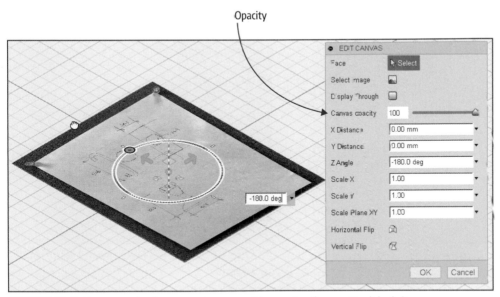

FIGURE 10-5: You can make the canvas transparent by moving the opacity slider left.

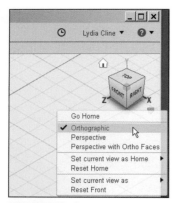

FIGURE 10-6: Set the workspace to Orthographic.

Set the workspace to Orthographic (Figure 10-6) to make modeling easier. We'll mostly work with the ViewCube turned to one side.

MODEL THE CLIP

The clip has three parts: a bracket that holds the phone, a ball joint, and a stem that holds the ball joint and bracket together.

Trace the Bracket Sketch

Draw two concentric circles as shown in Figure 10-7. Then copy them. The sketches must be in Edit mode to copy, so if you clicked Stop Sketch after drawing them, drag a window around them to select, right-click, and choose Edit Sketch (Figure 10-8).

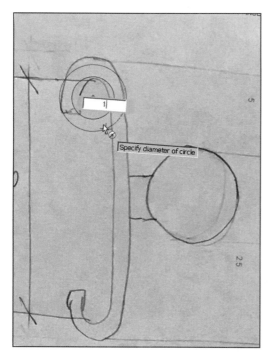

FIGURE 10-7: Trace the top curve with two concentric circles.

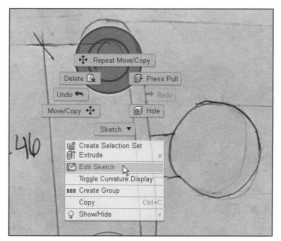

FIGURE 10-8: Enter Edit Sketch mode if you're not already in it.

Right-click the selected circles and choose Copy. Click the mouse onto the workspace; then right-click and choose Paste. Arrows will appear over the circles; drag them to move the copy to the bottom of the sketch (Figure 10-9). Then draw two vertical lines connecting the pairs of circles as shown in Figure 10-10.

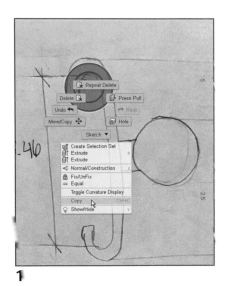

1

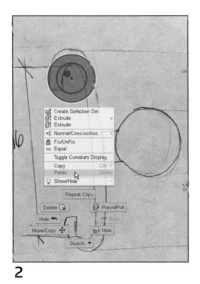

2

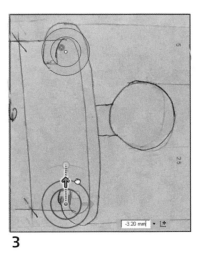

3

FIGURE 10-9: Copy the circles and move them into place.

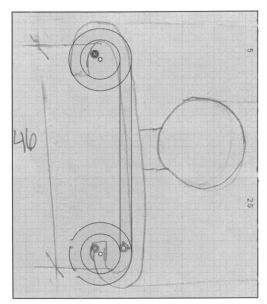

FIGURE 10-10: Draw two vertical lines.

Edit the Traced Bracket Sketch

Draw horizontal lines through the circles' centers to serve as trim lines. Use the snap function to make sure the lines are tangent to the circles (check the Snap To Grid box in the Snap And Grid icon at the bottom of the screen). Trim off the bottom half of the top circle and the top half of the bottom circle. Figure 10-11 shows how the top circle should look when finished.

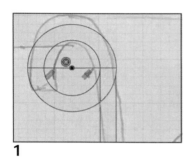 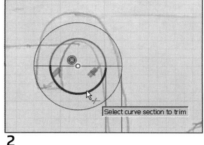 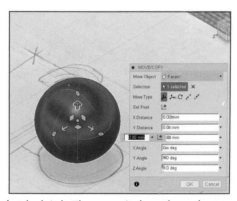

1   2   3

FIGURE 10-11: Trim the circles.

Make a Ball Joint

We'll use a sphere the size of the hand-sketched circle as a ball joint. After it is made, the sphere will probably be half above and half below the grid. Move it higher by right-clicking to select it and choosing Move/ Copy (Figure 10-12). Next, copy the sphere by selecting it through its Browser entry, right-clicking it, choosing Move/Copy, and selecting the Create Copy box (Figure 10-13). We'll need that copy later.

 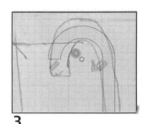

FIGURE 10-12: Make a sphere the size of the hand-sketched circle. Then move it above the workspace.

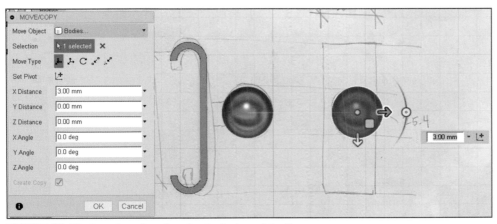

FIGURE 10-13: Copy the sphere by selecting it through its Browser entry.

Extrude the Bracket Sketch

Select and right-click the clip. Choose Press Pull and extrude the sketch up. I extruded it up 1.70 mm (Figure 10-14).

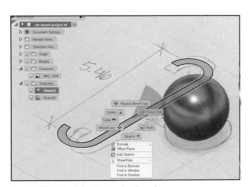 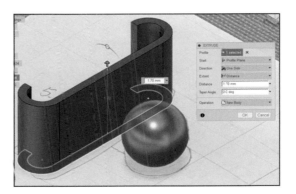

FIGURE 10-14: Extrude the clip up.

Make a Stem

Sketch a circle centered on the clip. To do this, first draw a line down the clip's center (hover the mouse along an edge until the midpoint [triangle] snap point symbol appears). Then draw a circle anywhere on the clip; it should be smaller than the ball joint. Use the midpoint constraint to snap the center of the circle to the center of the line (Figure 10-15).

Then delete the line sketch. Select the circle, right-click, and press-pull it forward to make the stem. Move the sphere onto it (Figure 10-16).

Alternatively, you can create the stem by going to the front view and moving the sphere to align its center with the stem circle sketch's center. Then do the same from the top view and extrude the circle.

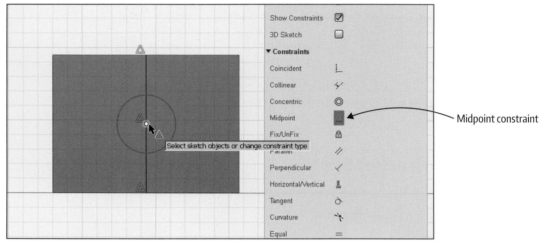

FIGURE 10-15: Center the circle on a line with the midpoint constraint.

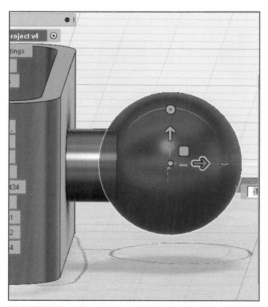

FIGURE 10-16: Extrude the circle sketch and center the sphere on it.

Round the Bracket's Edges

Select all the edges at once by clicking them while pressing and holding Shift. Then right-click one selected line and choose Fillet. Drag the arrow to curve the edge (Figure 10-17).

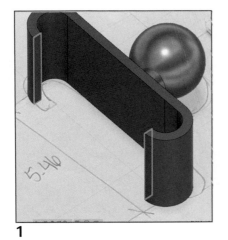
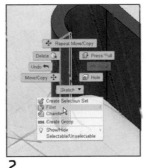

1   2   3

FIGURE 10-17: Fillet the edges.

Combine All the Clip's Parts

When you extruded the stem, it should have joined with the bracket. But if for some reason they remain as two separate parts, merge them together, a requirement for being 3D printable. Click Combine → Join. First combine the sphere with the stem (Figure 10-18), and then combine the sphere and stem with the bracket.

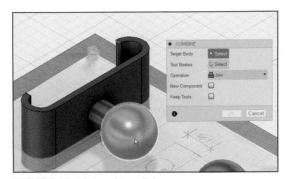
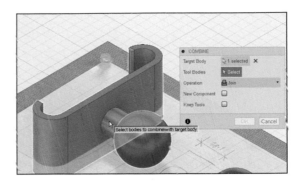

FIGURE 10-18: Combine all the parts.

MODEL THE BASE

Again, set the workspace to Orthographic, if it isn't already, to make modeling easier. The base consists of a tapered form, stem, and socket.

Make a Tapered Form

Draw two circles over the hand-sketched circles, with proportions as shown in Figure 10-19. Then select the small circle, right-click, and choose Move to move it up. Draw a selection window around the large circle, right-click, choose Press Pull, and extrude it down (Figure 10-20).

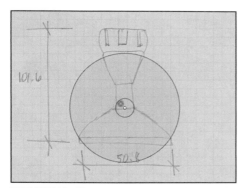 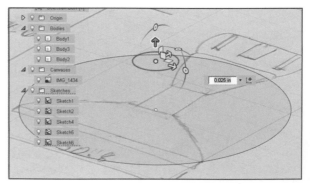

FIGURE 10-19: Draw two circles and move the smaller one up.

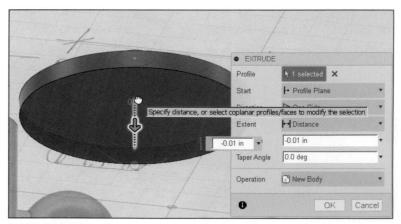

FIGURE 10-20: Select the large circle and extrude it down.

Now we'll loft between the circles. Select both sketches (you might need to turn on their light bulbs to see them). You must select the whole circle, not just their outlines. You also need to select them one at a time, so don't drag a selection window around them. Click their browser entries (press and hold Shift) or click inside the circle sketches. Then choose Create → Loft (Figure 10-21).

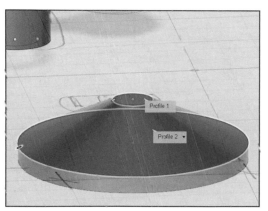

FIGURE 10-21: Select each circle individually and loft them.

Make the Stem

Sketch a circle on top of the tapered form. Select it, right-click, choose Press Pull, extrude it up into a cylinder (Figure 10-22), and click OK. Then sketch another circle on top of that cylinder and extrude it up. But before clicking OK, drag the button at the top to taper the stem outward (Figure 10-23).

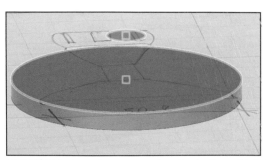

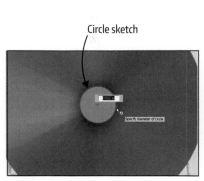

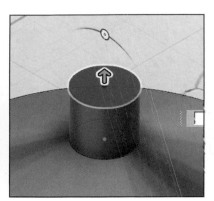

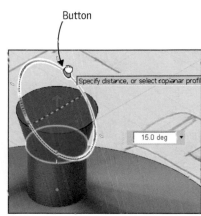

FIGURE 10-22: Sketch a circle on top of the base and extrude it up.

FIGURE 10-23: Extrude a second cylinder and taper its top.

Make the Socket

Turn the second sphere that you made earlier into a socket to hold the clip's ball joint. You might want to move the sphere up off the base a bit to make this operation easier. Draw a horizontal line across its middle—click one of the origin's vertical planes to start that operation. The line doesn't have to touch the sphere; it just needs to be centrally located on it. Then choose Modify → Split Body. In the dialog box, click the Body To Split button and then click the sphere. Next, click the Splitting Tool button and select the line. Click OK, and the sphere will be cut in half (Figure 10-24).

Select the top half through its Browser entry, right-click that entry, and choose Remove. This will remove the top from the timeline from this point forward (deleting the top would cause problems). Select the flat top as shown in Figure 10-25. Right-click, choose Shell, and drag the arrow a few millimeters to give it thickness.

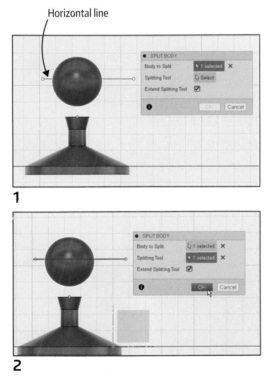

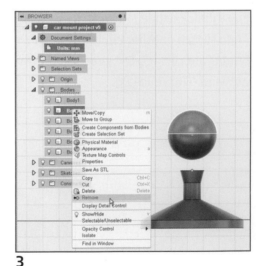

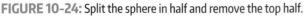

FIGURE 10-24: Split the sphere in half and remove the top half.

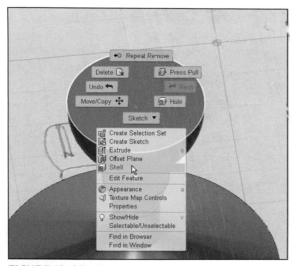

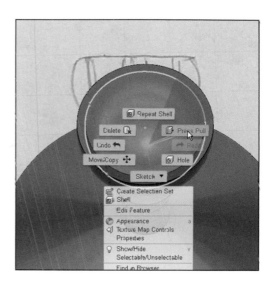

FIGURE 10-25: Shell the sphere.

MAKE THE SOCKET HOLES

Draw a circle that is slightly bigger than the cylinder on the clip's handle. Add a 3-point rectangle and trim the overlapping parts. The result should be the shape shown in Figure 10-26.

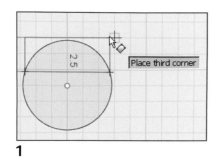

1

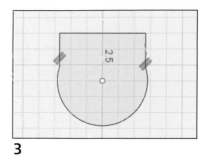

2

3

FIGURE 10-26: Sketch this shape with a circle and 3-point rectangle.

Enter Edit Sketch mode if you're not in it. Drag a selection window around the sketch, right-click, and choose Copy. Click the work plane; then right-click and choose Paste. Drag the copy off the original with the arrows and then rotate it 90° (Figure 10-27).

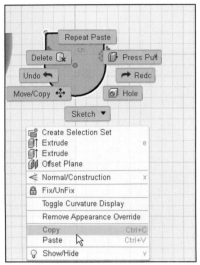

1 Copy

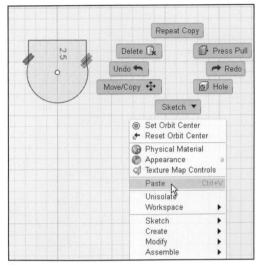

2 Click the work plane and paste

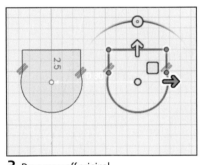

3 Drag copy off original

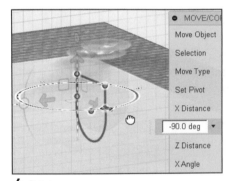

4 Rotate copy

FIGURE 10-27: Copy and rotate the sketch.

POSITION AND EXTRUDE THE SOCKET HOLE SKETCHES

Position the two socket hole sketches as shown in Figure 10-28 and extrude them through the shelled half-sphere. You can use the center point of the cylinder and the center point of the sketch to align them. Then, select the socket through its Browser entry, click Modify → Scale, and scale it a bit larger than the ball joint (Figure 10-29). Alternatively, change the scaling point to be the center of the stem (use the center of the circle that was used to extrude it). Then select the sketches, right-click, and choose Show/Hide to make them disappear.

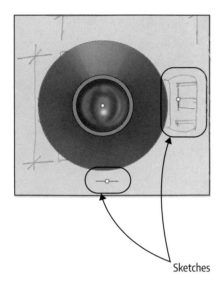

Sketches

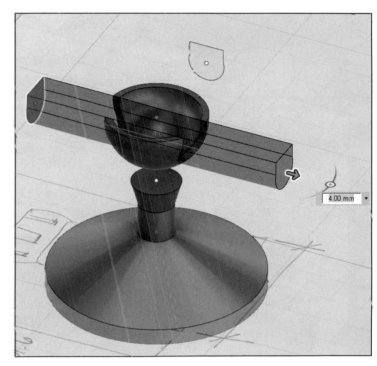

FIGURE 10-28: Position and extrude the sketches.

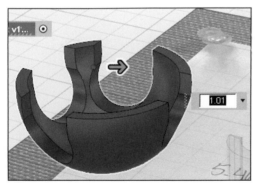

FIGURE 10-29: Scale the socket a bit larger than the ball joint.

Align and Join the Socket and Stem

If the socket and stem aren't already joined, you'll need to join them to make the base 3D printable. First, align them. Choose Modify → Align, click the socket, and then click the stem (Figure 10-30). The two parts will snap together. Then finesse the socket's position by moving it. Select it through its Browser entry, right-click, and move it (Figure 10-31). Using Modify → Combine/Join, combine the stem and socket (Figure 10-32).

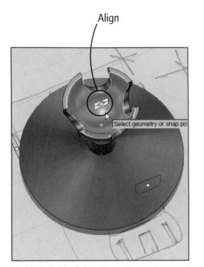
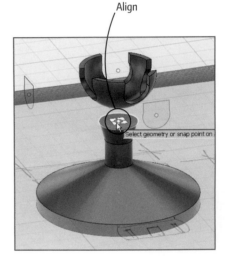

FIGURE 10-30: Align the socket and stem.

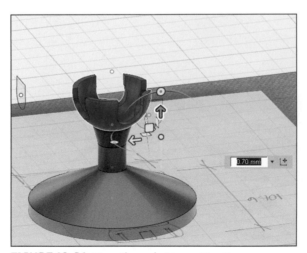

FIGURE 10-31: Move the socket to position it.

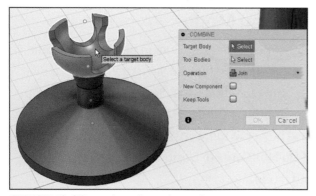
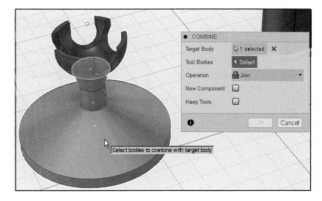

FIGURE 10-32: Combine the socket and stem.

SCALE THE CLIP AND BASE

We're going to scale the clip and base proportional to the width of the bracket. To do this, we need to know the bracket's current size and the size we want it to be.

Measure the Bracket's Current Size

Set the workspace to Orthographic mode, if it's not a ready. Then choose Inspect → Measure, and click the bracket's interior endpoints (Figure 10-33). The dialog box shows that distance as 3.10 mm.

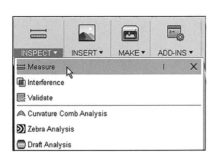

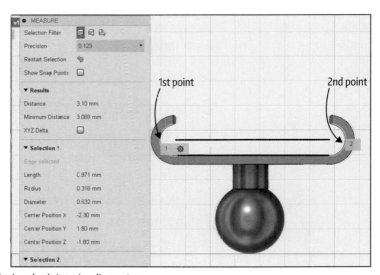

FIGURE 10-33: Click the Measure tool onto the bracket's interior dimensions.

Enter Direct Modeling Mode

Scaling in direct modeling mode is easier than in parametric mode. However, direct mode loses the timeline, so save the file with a different name in case you want to edit it later. Save the project, if you haven't already done so, and then choose File ➔ Save As. Type a new name, and click the Save button (Figure 10-34). Then right-click the title name and choose Do Not Capture Design History (Figure 10-35) to enter direct modeling mode.

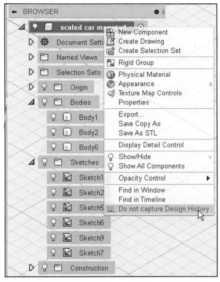

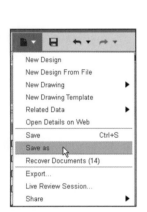

FIGURE 10-34: Save the file under a new name.

FIGURE 10-35: Choose Do Not Capture Design History.

Enter a Scale Equation

Drag a selection window around both the clip and stand to select them. Choose Modify ➔ Scale, and in the dialog box there will be a Scale Factor text field. For my project, I typed **5.46/3.10** (Figure 10-36) because 5.46 mm is the distance between endpoints that I want and 3.10 is the distance that I have. Click OK, and both parts will resize. Verify the new size is what you want by clicking the Measure tool onto the endpoints (Figure 10-37).

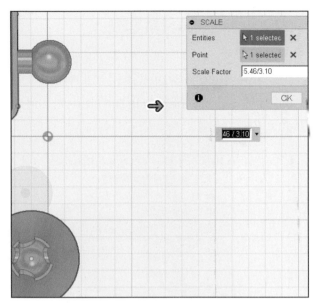

FIGURE 10-36: Scale the parts with an equation.

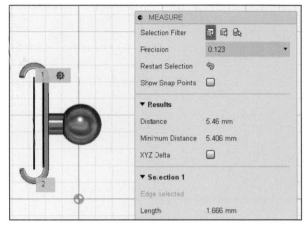

FIGURE 10-37: The clip now measures 5.46 mm.

Figure 10-38 shows the two finished parts.

FIGURE 10-38: The car mount's clip and stand

SAVE THE PARTS SEPARATELY AS STL FILES

Select the clip, right-click, and choose Show/Hide to hide it. Then save the stand as an STL by right-clicking the title in the Browser and choosing Save As STL (Figure 10-39).

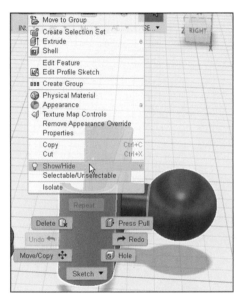 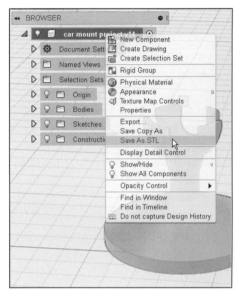

FIGURE 10-39: Hide the clip and save the stand as an STL file.

Find the clip's entry under Bodies in the Browser and click its light bulb to turn it yellow, which unhides it (Figure 10-40). Then hide the stand by clicking its light bulb to turn it gray and export the clip as an STL. You can now import both STL files into slicing software for 3D printing.

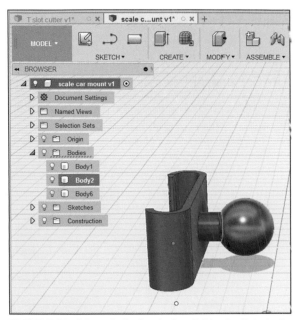

FIGURE 10-40: Unhide the clip by clicking its light bulb.

Print this project in ABS or PETG to withstand heat. Since all printers and filament types have different accuracy levels, a prudent Maker will slice interlocking parts off the rest of the model and print them first. That way, you can ensure they work before printing the whole model. If they don't work, adjust the scaled size. Stick the final product to the dashboard with a double-sided adhesive gel pad.

MAKE AN ASSEMBLY

An assembly is a model whose components are shown in their relative positions in the finished product. The components are held together with joints that mimic realistic mechanical motion via an animation.

Both the base and clip currently are two separate pieces each, with a total of four bodies. Combine the base's stem and ball joint holder. Then combine the ball joint and bracket. Now there are two bodies. Right-click each body and choose Create Component From Bodies (Figure 10-41).

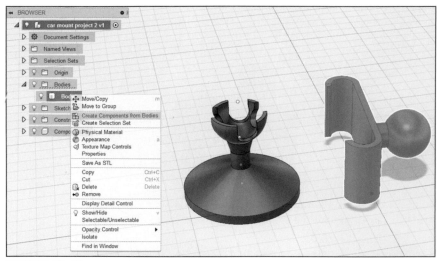

FIGURE 10-41: Turn the bodies into components.

Right-click the base's Browser entry and choose Ground (Figure 10-42). This is the stationary component; the other one will move relative to it.

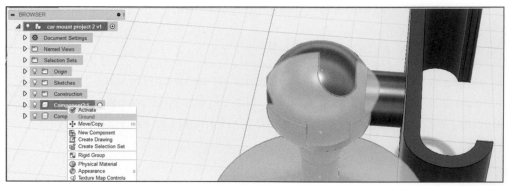

FIGURE 10-42: Ground one component.

Next, choose Assemble → Joint (Figure 10-43). A dialog box will appear; click the dropdown arrow to choose the appropriate joint. In this case it's the ball. Click both components to place the pin at the locations you want them to touch (Figure 10-44)

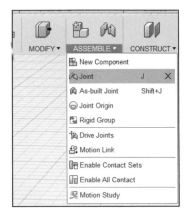

FIGURE 10-43: Click Assemble → Joint.

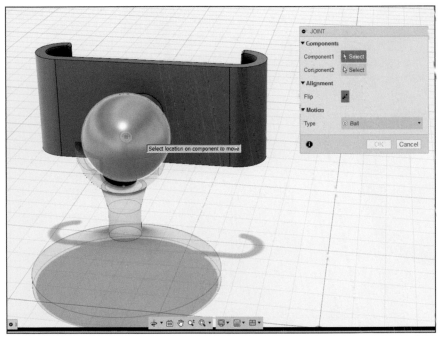

FIGURE 10-44: Choose the appropriate joint and click it onto the components.

The first surface you click will be the part that moves to the other part. The two parts will animate, showing you how the model works (Figure 10-45). You can also right-click the component's Browser entry and choose Animate Joint (Figure 10-46).

FIGURE 10-45: The final assembly

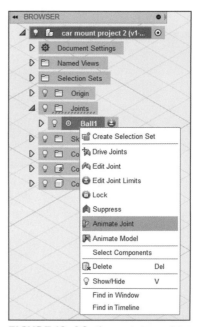

FIGURE 10-46: Choose Animate Joint.

11

2D LAYOUT

In this chapter, we'll turn the Chapter 9 planetary gear set into a sheet of orthographic, scaled drawings. Such documentation is needed for shop fabrication or to prepare drawings suitable for a CNC cutting machine.

CONVERT BODIES TO COMPONENTS

Open the gear model. Convert all bodies in the file into components. Although solid bodies can enter the drawing workspace unconverted, they lack full functionality there. Mesh, T-spline, and surface bodies will not enter the drawing workspace at all.

ENTER THE DRAWING WORKSPACE

Click the Workspace menu and then select Drawing → From Design. Figure 11-1 shows two ways to do this. If the Fusion file was never saved, you'll be asked to do so now. Then a dialog box will appear (Figure 11-2). Note the Full Assembly option. If you uncheck it, you can select individual components instead of bringing the whole file in. Hold down the Shift or Ctrl key to make multiple selections.

FIGURE 11-1: There are two ways to enter the Drawing workspace.

You can also choose a template from scratch or browse for one; choose the ISO or ASME standard, choose inches or millimeter units, and choose the sheet size. We'll use the default size. The sheet size can be changed after you create a drawing, but the standard and units cannot. Then click OK to enter the Drawing workspace. When you do, a tab for this drawing will appear at the top of the screen.

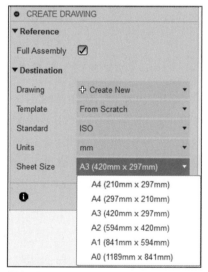

FIGURE 11-2: Choose options.

BASE VIEWS

When you enter the workspace, a *base view*, also called the *parent view*, of the gear will appear. This is a 2D view derived directly from the model. There will also be a dialog box with options for, among other things, orientation, style, and scale. Here you choose to display the base view from the top, front, or side, or as an isometric (a 3D view). You can also choose the scale (Figure 11-3). Shown is a 1:5 scale, which can be changed on each base view, or changed on this view later.

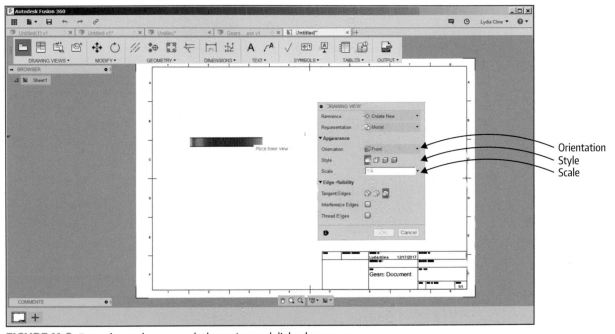

FIGURE 11-3: Enter the workspace, and a base view and dialog box appear.

Create Multiple Base Views

To bring in another base view, go to the Drawing Views menu. Click the dropdown arrow, click Base View, and bring a second view into the workspace (Figure 11-4).

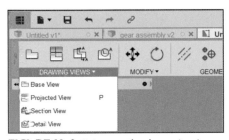

FIGURE 11-4: Bring another base view in through the Drawing Views menu.

Edit a Base View

When you click a base view, a dashed blue box surrounds it, indicating it's highlighted and can be edited. Right-click to bring up a context menu (Figure 11-5). Click Edit View to open the Drawing View box. I clicked the Orientation dropdown arrow and then clicked TOP. The highlighted view changed from the default side view to a top view. Figure 11-6 shows it as an isometric view. To delete it, or any other view, just highlight it and press the Delete key.

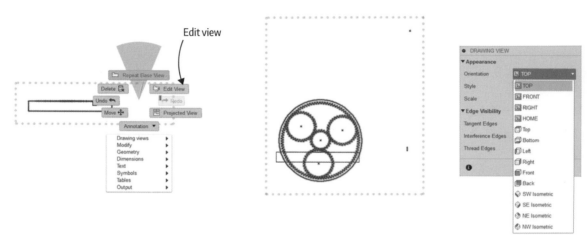

FIGURE 11-5: Changing a base view from the default side view to a top view

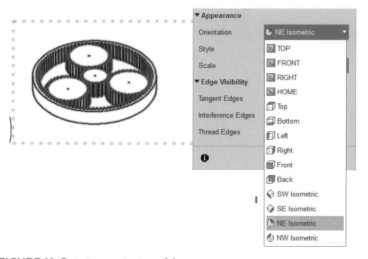

FIGURE 11-6: An isometric view of the gear

In Figure 11-5's middle graphic you can see that the top view of the gear overlaps the side view. Click that side view to bring up its blue editing box. A small gray square will appear within the box; drag the square to relocate the view (Figure 11-7).

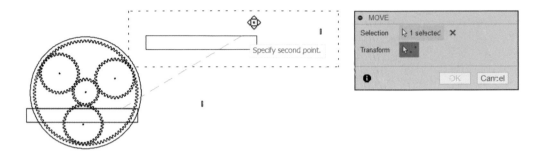

FIGURE 11-7: Move a view by dragging it when the blue editing box is open.

Projected View

A projected view is one generated from a base view, and it inherits its properties from the base view it's projected from. To project one view from another, choose Drawing Views ➙ Projected View. Then click the base view and the location you want to place the projected view (Figure 11-8). The projected view will automatically position itself depending on where it is in relation to the base view.

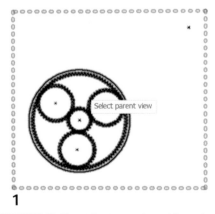

1

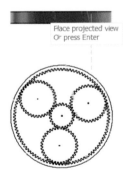

2

FIGURE 11-8: A side view projected from the top view

After placing the projected view, you can change its properties by double-clicking it. A View Properties dialog box will appear; choose options. When you click OK, the projected view will automatically update.

Section View

A section view is a slice through the whole model, which allows you to see inside it. To make one, choose Drawing Views → Section View. Click the plane view or parent face to start, click two points on the plane or face where you want to make the slice, and then press Enter (Figure 11-9). Select the newly created section and drag to place (Figure 11-10).

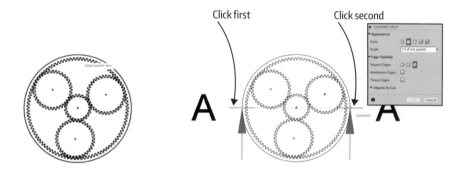

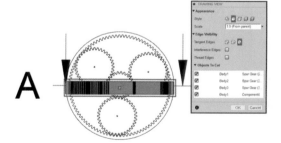

FIGURE 11-9: Cutting a section view

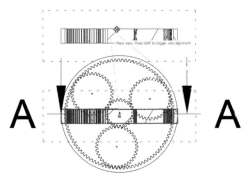

FIGURE 11-10: Drag to relocate the section view.

CENTER LINE

To put a center line through a view, choose Geometry → Centerline.
Then click two straight edges. A center line appears between them. Drag
the arrows on each end to make the center line longer (Figure 11-11). If a
center line is too short, it won't appear as a center line type.

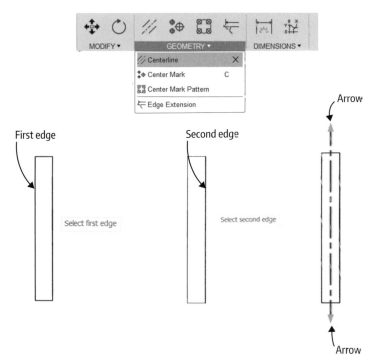

FIGURE 11-11: Putting a center line in a base view

DIMENSIONS

Dimensions are notes on lines that describe size. To add dimensions to a view, click Dimensions. There are many options in this menu; we'll choose Linear Dimension (Figure 11-12).

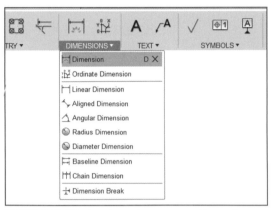

FIGURE 11-12: The Dimension menu and its options

Click the item to be dimensioned. You can click points, edges, or existing dimensions, or click two points. Then click to place the dimension line (Figure 11-13), which also finishes the operation. Double-click the dimension line after you've set it to bring up a preferences dialog box.

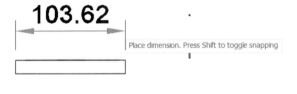

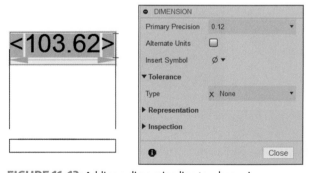

FIGURE 11-13: Adding a dimension line to a base view

TEXT AND LEADER TEXT

To add text, click Text (Figure 11-14). The Text option places text anywhere on the workspace. Click to set the corners of the text box, and a dialog box appears with options for font and size. Type the text and then either click outside the text box to finish or click Close in the dialog box (Figure 11-15).

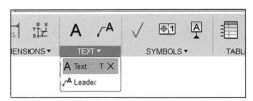

FIGURE 11-14: The Text menu

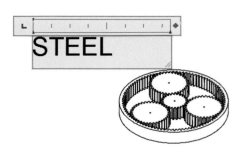

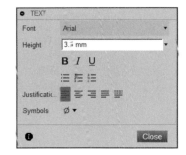

FIGURE 11-15: Adding text to the workspace

The Leader option attaches an arrow to the text. Click the arrowhead location, click the text location, and then type the text. Click outside the text box to finish (Figure 11-16).

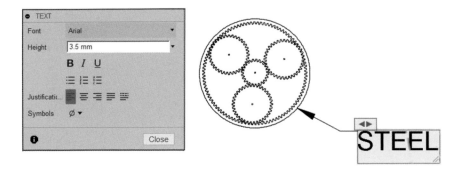

FIGURE 11-16: Adding leader text to a base view

CALLOUTS

To add a callout (an identification symbol) to a base view, choose
Symbols → Datum Identifier (Figure 11-17). Then click the place you
want to add it and type in the text field (Figure 11-18).

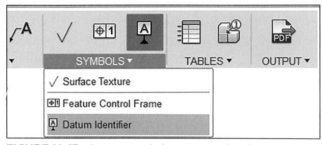

FIGURE 11-17: Choosing Symbols → Datum Identifier

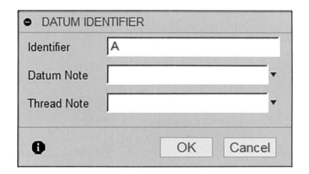

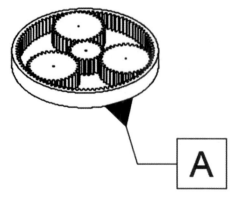

FIGURE 11-18: Adding a callout to a base view

TABLES

A table is a parts list for a single component, assembly, sheet metal flat pattern, or animation reference (a drawing created from an animation). Choose Tables → Table (Figure 11-19). A table will immediately generate for the whole assembly; drag to place it where you want (Figure 11-20). Select the table and drag gray squares in each column to adjust their width. To move the whole table, drag a selection window around it and click Modify → Move at the top of the screen.

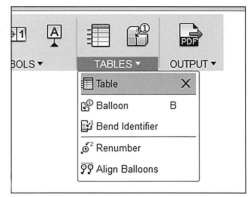

FIGURE 11-19: Choosing Tables → Table

Parts List				
Item	Qty	Part Number	Description	Material
1	1	Spur Gear (96 teeth)	Spur Gear; Module: 1.0; Num Teeth: 96; Pressure Angle: 20.0; Backlash: 0.00 mm	Steel
2	1	Component4		Steel
3	1	Spur Gear (24 teeth)	Spur Gear; Module: 1.0; Num Teeth: 24; Pressure Angle: 20.0; Backlash: 0.00 mm	Steel
4	1	Spur Gear (36 teeth)	Spur Gear; Module: 1.0; Num Teeth: 36; Pressure Angle: 20.0 Backlash: 0.00 mm	Steel

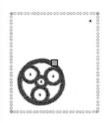

FIGURE 11-20: Automatically generate a table for the whole assembly.

You can generate balloon callouts to coordinate with the table by choosing Table → Balloon and then clicking a specific component (Figure 11-21).

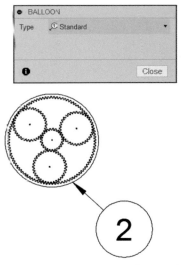

FIGURE 11-21: A balloon callout made from the Tables menu

TITLE BLOCK

The default sheet we chose has a built-in title block. Click that title block to bring up a window with text fields into which you can type your specific information (Figure 11-22).

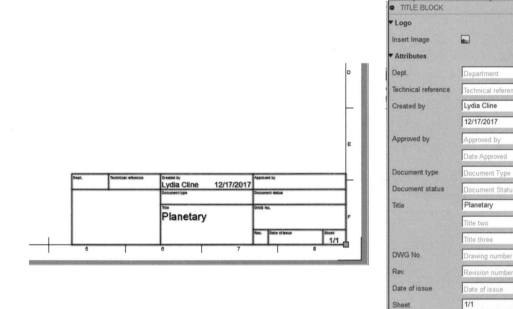

FIGURE 11-22: Click the title block to open this window.

EXPORT THE SHEET

Once the sheet looks as you want it, click Output (Figure 11-23) and save it as a PDF, DWG, or CSV file. The first two formats are files needed to manufacture the part. If your goal is to cut one gear out on a CNC machine, just send a model of one gear to the Drawing workspace and set it as a top view. The third format is a plain text parts list, which you can see in Figure 11-23. To exit, just close the tab at the top of the screen (Figure 11-24). The drawing file will close and a thumbnail will appear in Fusion's Data panel alongside all the model files.

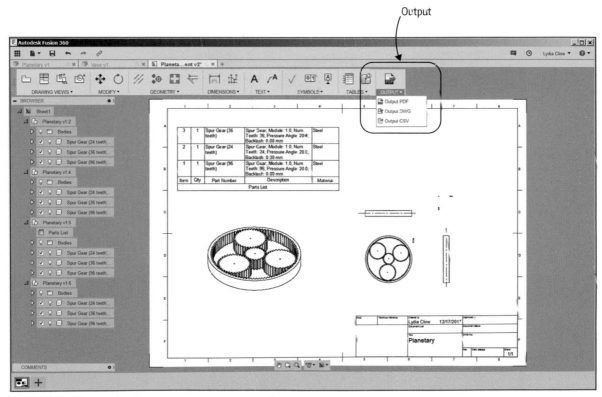

FIGURE 11-23: Save the sheet as a PDF, DWG, or CSV file.

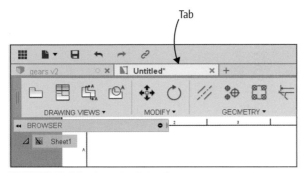

FIGURE 11-24: Close the file's tab to exit.

12

SLICER FOR FUSION

In this chapter, we'll use Autodesk's Slicer for Fusion app to make
2D patterns from a model.

Slicer for Fusion is not a "slicer" in the way that the word is used for
software that turns an STL file into 3D-printable g-code. It is a program
that cuts patterns that can be exported as PDF and DXF files and used
as templates to cut flat sheets with manual tools (such as a band saw
or jigsaw) or CNC cutting tools (such as a laser or router). Typical Slicer
projects include puzzles, terrain models (Figure 12-1), folded panels,
sculptures, furniture, crafts, and prototypes.

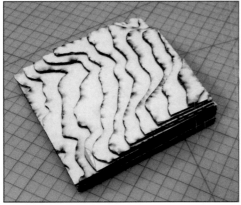

FIGURE 12-1: This terrain model's parts were cut
from a 3D model using Slicer.

We're going to cut one of the chairs modeled in Chapter 7. So open that file or any other file of your choice. The model needs to be a component, so if it's a body, convert it.

DOWNLOAD, INSTALL, AND LINK

Download Slicer for Fusion from the Autodesk App Store (Figure 12-2) and install it. You can access the Autodesk App Store through the Add-Ins menu or directly at **apps.autodesk.com**. Then launch Slicer directly by clicking its desktop icon on a PC (Figure 12-3), from the applications folder on a Mac, or from inside Fusion, where it will be an entry under the Make menu once you link it there. Launching it directly is useful when you want to import a non-Fusion file into it. Launching it from inside Fusion is faster because the Fusion model is sent right to it.

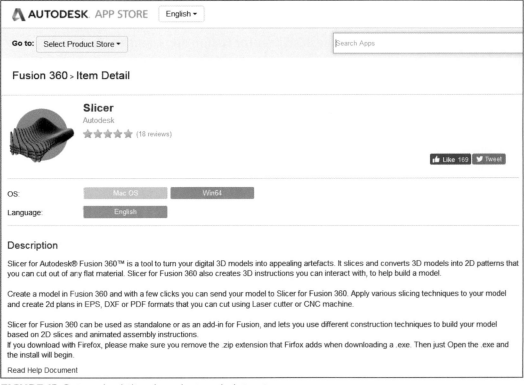

FIGURE 12-2: Download Slicer from the Autodesk App Store.

To launch Slicer from Fusion, you must link the two (Figure 12-4). Here's how:

FIGURE 12-3: The Slicer desktop icon

1. Choose Make → 3D Print.

2. Choose Print Utility → Custom.

3. Click the folder icon and navigate to Slicer's EXE file (C:\ Program Files (x86)\Autodesk\Slicer for Fusion 360\bin\ SlicerforFusion360.exe).

4. Click Open.

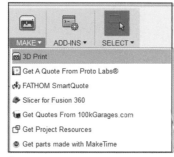

1 Click 3D Print

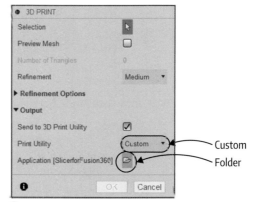

Custom

Folder

2 Click Custom and then the folder

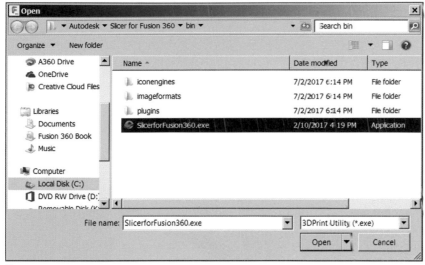

3 Navigate to the Slicer EXE file

FIGURE 12-4: Linking Slicer to Fusion

An entry for Slicer is now listed under the Make menu (Figure 12-5). If you don't see it, restart Fusion. Click the Slicer entry and a dialog box appears asking you to select one component to send to Slicer.

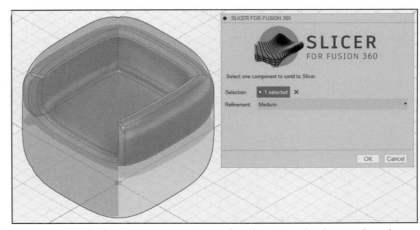

FIGURE 12-5: Click the Slicer entry under the Make menu and select one component to send to Slicer. Here the chair is selected.

TIP Slicer has no editing tools, so files sent to it should be finished products. If your file won't import into Slicer, or imports with missing parts, it may be due to flaws that make it unprintable, such as non-manifold edges, reversed normals, or holes. Repair it with analysis software such as Autodesk Netfabb or Meshmixer, or redo it in the original program.

THE INTERFACE

Once you send the file to Slicer, you'll see the interface in Figure 12-6. If the model entered at an odd angle, it's because the program it was imported from doesn't share the same coordinate system. Adjust it by clicking the Rotate icon on the left vertical panel (Figure 12-7), or reorient it in the original program.

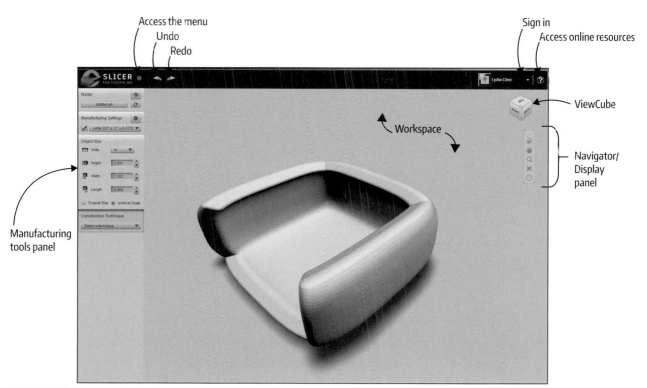

Access the menu
Undo
Redo
Sign in
Access online resources
ViewCube
Workspace
Navigator/
Display
panel
Manufacturing
tools panel

FIGURE 12-6: The Slicer interface and imported chair

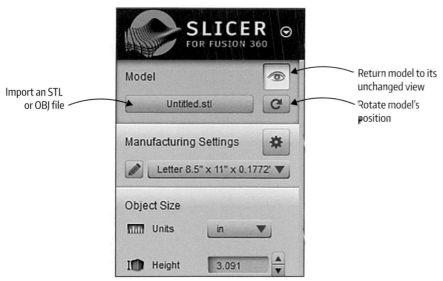

Import an STL
or OBJ file
Return model to its
unchanged view
Rotate model's
position

FIGURE 12-7: Adjust the file's position with the Rotate icon.

FIGURE 12-8: Click the dropdown arrow in the menu bar to see this submenu.

FIGURE 12-9: Links to online resources are on the menu bar.

FIGURE 12-10: Use the Import button to bring in a file.

There's a workspace, a menu bar at the top, and a vertical panel on the left that contains manufacturing settings and construction techniques. The menu bar has a dropdown arrow that accesses a submenu, Undo/Redo arrows (Figure 12-8), and links to online resources (Figure 12-9). There is also a ViewCube and Navigation/Display panel. Note that if you import a file directly into Slicer using the desktop icon, the interface will initially look empty (Figure 12-10). Once you import a file, the interface will look as just described.

The menu bar's submenu contains these functions:

▶ **New:** This closes the current file and makes the Import button appear.

▶ **Open Example Shapes:** This has premade files to experiment with.

▶ **Open:** This brings up a browser window from which you can open a project from your online Autodesk account (called From Fusion Team) or your computer.

▶ **Save:** Click here to save a 3DMK (Slicer) file to your online Autodesk account or your computer.

▶ **Save A Copy:** Click here to save a copy of the 3DMK file to your online Autodesk account. It will become the current file.

▶ **Export Mesh:** Turn the 3DMK file into an STL or OBJ file.

▶ **Exit:** This closes the file and exits the program.

NAVIGATING THE INTERFACE

You can move around the interface in three ways: with the ViewCube, the Navigation bar (Figure 12-11), and the mouse.

- ▶ **ViewCube:** This shows the model's orientation on the workspace. Click it and drag to rotate. The model will rotate with it. Click the ViewCube's sides to see the model orthographically—that is, as a 2D top, front, or side view. Hover the cursor over it to make a house appear; click the house to restore the file to the default position.

- ▶ **Navigation/Display panel:** This panel contains tools for orbiting, panning, zooming, and displaying the model.

  - ‣ **Orbit:** Hold down the right mouse button to orbit at any angle around the model.

  - ‣ **Pan:** Slide the model around the workspace.

  - ‣ **Zoom:** Click and drag the cursor up and down for a view that's closer to or farther from the model.

  - ‣ **Fit:** Click this icon to make the model zoom in to fill the screen.

  - ‣ **Display:** Click to switch between perspective (3D) and orthographic (2D) views.

- ▶ **Mouse:** Right-click anywhere on the screen and drag the cursor to orbit around the model. Press the scroll wheel down to drag the cursor to pan (slide) the model around the workspace. Roll the scroll wheel up and down to zoom in and out.

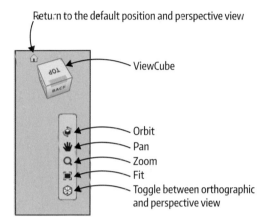

Return to the default position and perspective view

ViewCube

Orbit
Pan
Zoom
Fit
Toggle between orthographic and perspective view

FIGURE 12-11: The ViewCube and Navigation bar

SHEET SIZE AND MANUFACTURING SETTINGS

Select the sheet size the project will be cut from. In the Manufacturing Settings box (Figure 12-12), click the dropdown menu and scroll through the choices. Presets P1, P2, and P3 at the bottom are useful if you'll send the project to an online CNC cutting service.

If your material is different from any of the ones listed, click the pencil icon to access a screen that shows all settings choices. There are graphics at the bottom for adding your own presets, useful for nonstandard materials. Click the plus sign to add a new preset. To make a variation of a preset, click that preset to highlight it, and then click the double-plus sign to make a duplicate. Enter the new settings. To delete a preset, highlight it and click the minus sign. You can change or delete only presets that you create; you can't change the existing ones.

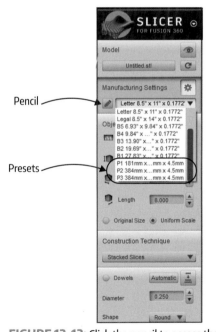
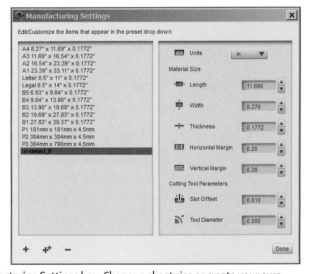

FIGURE 12-12: Click the pencil to access the Manufacturing Settings box. Choose a sheet size or create your own.

Now select the object's physical size (Figure 12-13). Choose its units and adjust the height, width, and length as needed. Clicking Uniform Scale will scale the file in all directions. When it's deselected, you can scale the model differently along the three axes. Clicking Original Size reverts the file back to the size at import. The larger the file is, the more slices that will be generated and the more sheets required.

CONSTRUCTION TECHNIQUE

At the bottom of the panel is the Construction Technique box (Figure 12-14). Click the dropdown arrow to see the options: Stacked Slices, Interlocked Slices, Curve, Radial Slices, Folded Panels, and 3D Slices.

Click each technique to see its effect; once you select it, the model will immediately update. Two-dimensional graphics of slices laid out on sheets will appear in the Cut Layout tab on the right (Figure 12-15). Each technique has at least one option unique to it, but most options overlap between them. As you finesse the options, the slices automatically update. (Slices and parts are not the same. A slice can consist of multiple parts.) To return to a view of the uncut model, click the eye icon at the top of the panel.

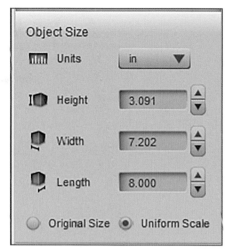

FIGURE 12-13: Choose the model's units and size.

FIGURE 12-14: The Construction Technique box

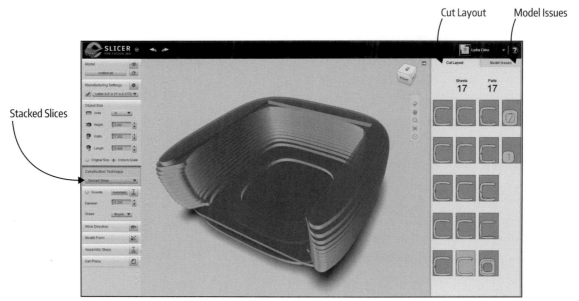

Cut Layout

Model Issues

Stacked Slices

FIGURE 12-15: The Stacked Slices technique applied to the chair. It applies cross-sectional slices.

THE SLICER PROCESS

Let's go through the whole Slicer process, from choosing the technique to printing the file. We'll use the Stacked Slices technique.

Apply a Manufacturing Technique

Click the Stacked Slices option. This makes cross-sectional slices to glue and stack. The Dowels option creates rods for aligning and holding the slices together (Figure 12-16). You can choose their size, location, and shape. Move a rod by highlighting and dragging it; create one by clicking a slice; delete it by highlighting it and pressing Delete.

Dowel

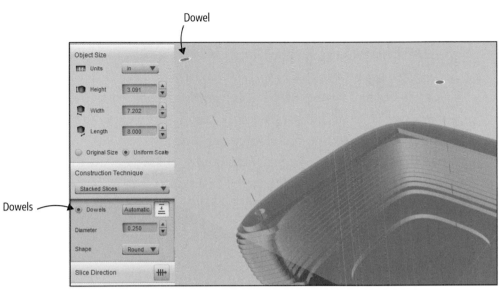

FIGURE 12-16: Click a rod to highlight it and adjust its location by dragging.

Change the slice direction (Figure 12-17) by dragging the blue handle. The handle can be dragged only when it's visible (dark blue), not hidden (muted). Select the Model Issues tab to see problems such as unconnected parts; they'll show up in blue. Our model isn't showing any problems (Figure 12-18).

Handle

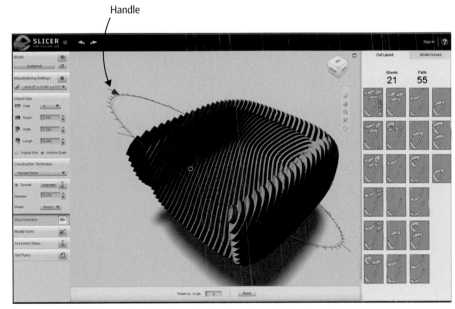

FIGURE 12-17: Change the slice direction by dragging the blue handle.

FIGURE 12-18: Select Model Issues to see problems. No problems appear here.

Optional: Modify Form

Even if everything looks fine, problems may appear when exporting the model to a different file format. In that case, you can click the Modify Form icon. Three buttons will appear at the bottom of the screen: Hollow, Thicken, and Shrinkwrap (Figure 12-19). Hollow shells out the model, reducing the amount of material needed to build it. Thicken widens thin cutouts. Shrinkwrap smooths details that are too fine to cut out. It also closes holes, useful on models made by scanning. Select an option, adjust its slider, and click Done.

FIGURE 12-19: Use the Modify Form option to fix problems that show up after the file has been exported.

Shrinkwrap affects the whole model, not just a problem slice. So the end product may not look exactly like the model. Additionally, know your cutter's limitations; for instance, a router can't make square internal corners without a v-bit. Appropriate clearance between connecting parts is also needed.

Assembly Steps

Click the Assembly Steps icon and choose material (Figure 12-20) to display on the model. Drag the mouse along the slider at the bottom of the screen to view an animation putting the parts together. To see an enlargement of an assembly sheet, click it in the Assembly Reference panel on the right (Figure 12-21). Move the sheet by clicking and dragging; scroll the mouse wheel to zoom in on a part; click the X in the upper-right corner to return to the model.

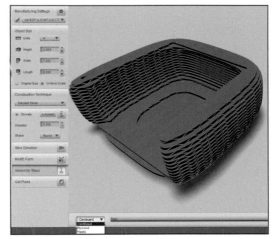

FIGURE 12-20: Click Assembly Steps to display a material and see an animation of how to put the parts together.

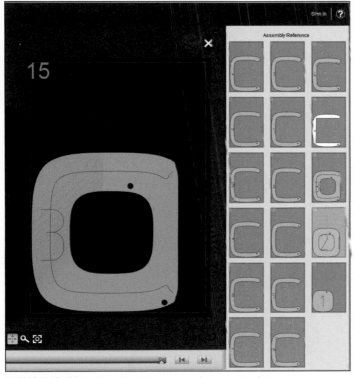

FIGURE 12-21: Click a sheet to see an enlargement of it.

On the Cut Layout tab, the graphics show how many sheets of material are needed, how many slices and parts there are, and where problems are. The parts are arranged to use as much of each sheet as possible, and all are oriented the same direction. Currently, there is no other way to arrange them. If you want to change the arrangement, such as to change the direction of a wood grain, import the file into a vector program such as Inkscape or Illustrator and change it there. Alternatively, your cutting machine's slicing software might be able to do that.

> **TIP** Hyphenated labels on the parts, such as *Z-3-2*, describe the part's axis (Z), the slice number (3), and the part number (2).

EXPORT THE FILE

Click the Get Plans icon, choose the format (EPS, PDF, DXF) and units, and click Print. A dialog box will appear with the choice to print the current sheet or all sheets on separate pages (Figure 12-22). To get DXF and EPS files, click the Export To My Computer or the Export To Fusion Team button (the latter saves it to your online Autodesk account).

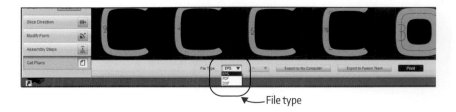

File type

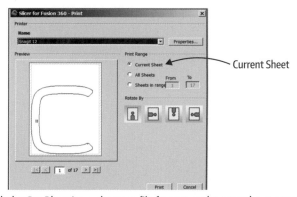

Current Sheet

FIGURE 12-22: Click the Get Plans icon, choose a file format, and export the current page or all pages.

OTHER CONSTRUCTION TECHNIQUES

Let's look at other construction techniques and their options now.

Interlocked Slices

This slices the model into two stacks of slotted parts that lock together in a grid. It uses less material than the Stacked Slices technique. Problem slices show up in red (Figure 12-23), which you can fix by adjusting the options (Figure 12-24). Those options include slice distribution, notches, and relief. Distribute the slides by count, distance, or a custom setting. The First Axis field sets the number of slices in each direction, and the Second Axis field sets the distance between slices. Increase the numbers to make the model more detailed; decrease them to use less material. Any problem slices will be shown on the Cut Layout tab.

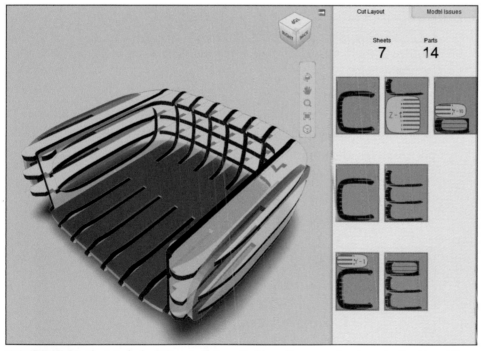

FIGURE 12-23: The Interlocked Slices technique. Problem slices are red and may be fixed by adding slices using the First Axis option.

FIGURE 12-24: Options for the Interlocked Slices technique

Click the horizontal bars to see icons that duplicate, delete, and evenly distribute the slices (Figure 12-25). Click and drag a slice to change its position. Click Evenly Distribute after moving individual slices to evenly space them. To add a slice, highlight an existing one and click the Add button. This creates a duplicate next to it that follows the model's contours. To remove a slice, highlight it and press Delete.

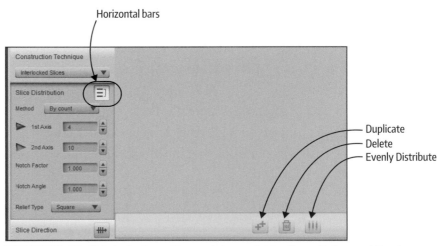

FIGURE 12-25: Click the horizontal bars to bring up the Duplicate, Delete, and Evenly Distribute buttons.

Large interlocked parts can twist, making assembly difficult or impossible. The default square cornered slots may also cause assembly problems, depending on the design. Use Notch Factor to flair the slot by a specific amount relative to its width; use Notch Angle to set the angle relative to the slot's direction. A 45° angle often works (Figure 12-26). Notches of any type must be thick enough to cut out.

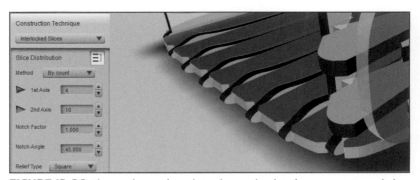

FIGURE 12-26: Change the notch angle to change the slots from square to angled.

Relief refers to the notch shape. The default relief is square, but you can choose a horizontal, vertical, or dog bone (which is a shape with no inside rounded corners). You can further edit the relief by specifying the tool diameter. Go back to the Manufacturing Settings box, make a new preset, and enter a number in the Tool Diameter field. If the Tool Diameter is 0, dog bones can't be generated. Alternatively, click the gear next to Manufacturing Options. A bar at the bottom will appear where you can change settings.

Curve

This technique cuts slices perpendicular to a curve (Figure 12-27). All the curves are on one plane. It works best on flowing, organic shapes, such as plants and animals. Options are the same as with the Interlocking Slices technique; additionally, you can bend the curve.

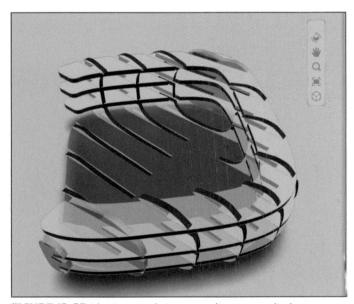

FIGURE 12-27: The Curve technique cuts slices perpendicular to a curve.

Radial Slices

This technique creates radiating slices from a central point (Figure 12-28). Use it on round, symmetrical objects.

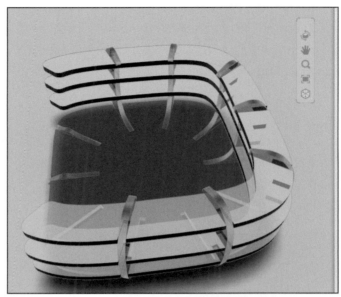

FIGURE 12-28: Slices in the Radial Slices technique emanate from a central point.

Folded Panels

This technique turns the model into 2D segments, or panels, of triangular meshes that you fold multiple times. Paper, cardboard, and sheet metal are common materials for this technique (Figure 12-29).

Figure 12-30 shows the default generated chair. It has a lot of red problem areas. There's also a message that the parts are too large for the sheets, so we need to choose a different sheet in the Manufacturing Settings box. Clicking the Split Panels option turns every face into a panel, which fixes them but generates hundreds of parts (Figure 12-31).

FIGURE 12-29: This store display is made with folded paper panels.

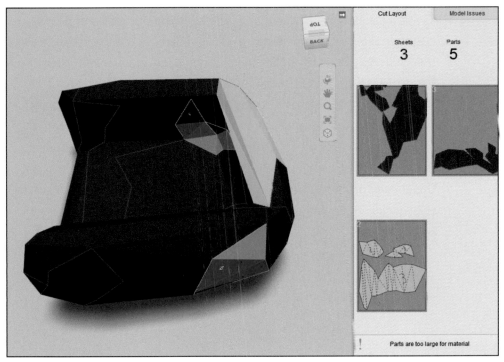

FIGURE 12-30: The default generated chair

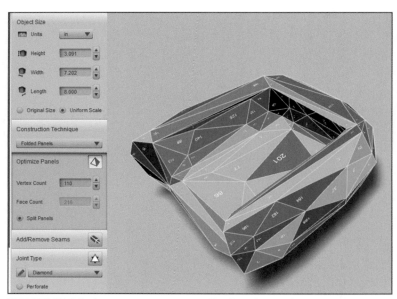

FIGURE 12-31: The many parts generated by the Split Panels option

FIGURE 12-32: The Export Mesh option preserves a copy of the model with its mesh intact.

You can decimate the model, which lowers the vertex and face count, making assembly easier. If you want to use the decimated model but preserve a copy of the file with its mesh intact, export it using the Export Mesh option in the menu bar (Figure 12-32).

Click the Add/Remove Seams icon and then click an edge to add fold lines that divide large panels into smaller ones or remove fold lines to combine small panels into bigger ones. You can also turn fold lines into perforation lines. The perforations will be the thickness of the chosen material. The Split Panels option will deselect when you do this; click it again when finished if your model becomes covered with problem areas.

Folded panels have joints (tabs) that connect the panels. There are 10 types (Figure 12-33). If a model has lots of problem areas, changing the joint type may help.

The joint types are:

- **Diamond:** Fold and glue/weld the triangular tabs.
- **Gear:** Fold and glue/weld the rectangular tabs (cut the dark areas out).
- **Laced:** Entwine the sheets together.
- **Multi-tab:** Fix the tabs together.
- **Puzzle:** Fit pieces together.
- **Rivet:** Pin tabs to one another.
- **Seam:** Sew tabs together (like a sewing pattern).
- **Tab:** Insert tabs into slots.
- **Ticked:** Connect tabs along the seams.
- **Tongue:** Insert a tab into slots.

FIGURE 12-33: Click Joint Type to access different options.

3D

This technique makes slices of the model that follow its form, as opposed to merely stepping them to create a form, as with the Stacked Slices technique (Figure 12-34).

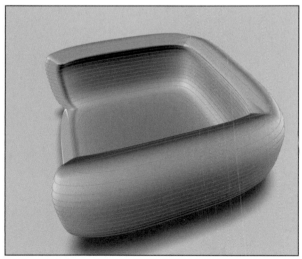

FIGURE 12-34: The 3D technique makes slices that follow the model's form.

SAVE AND EXPORT

You can save your work on your computer or to your online Fusion account; refer to Figure 12-8. You can also export the model as an STL or OBJ file to your desktop.

Additional Resources

Download Slicer: **https://apps.autodesk.com/en**

Ponoko, an online company that fabricates your designs: **www.ponoko.com**

13

THE CAM WORKSPACE

Fusion's computer-aided manufacturing (CAM) workspace creates g-code files for CNC cutting machines. It also exports DXF files for use as stencils for manual cutting. In this chapter, we'll generate toolpaths and an NC (numerical control) file for the emoji plaque modeled in Chapter 6.

WHAT IS G-CODE?

G-code is a programming language. It is used in CAM to control automated machine tools. Basically, it tells a machine how to make something: where to move, how fast to move, and what path to follow. The cutting tool is moved according to these instructions and cuts away material, leaving the finished workpiece. G-code is a type of NC code.

WHAT IS A TOOLPATH?

A toolpath tells the cutting machine what to cut, the direction to cut it, how fast to move, how many rough and finish passes to make, the distance for moving off and back onto the material, and other parameters. It's not a file on its own.

WHAT IS A CNC MACHINE?

A CNC machine is a waterjet, laser, lathe, or router that is controlled by computer, not manually. Routers use end mills, a rotating cylinder that looks like a drill bit but cuts both axially and laterally. A drill bit cuts only axially, creating cylindrical holes (Figure 13-1). End mills have hundreds of different tip designs for cutting different shapes (Figure 13-2).

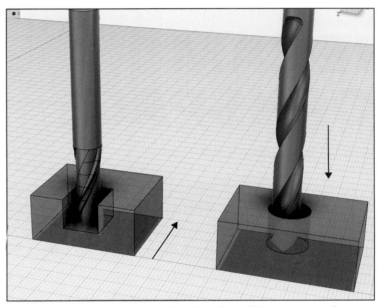

FIGURE 13-1: End mills cut laterally and axially. Drill bits just cut axially.

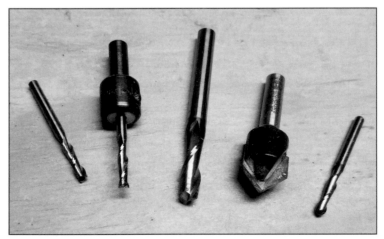

FIGURE 13-2: End mills have different tip designs for different kinds of cuts. (Photo courtesy of Paul Yanzick)

WHAT ARE 2D, 2.5D, AND 3D CUTTING?

The terms 2D, 2.5D, and 3D describe the cutter's movement along axes (an axis is a linear or circular direction of motion). Figure 13-3 shows examples of each. A 2D machine cuts in two directions (x and y), making a flat outline. A 2.5D machine cuts up and down as well (x, y, and z) so that you can add grooves or flat pockets. However, it can cut in only two directions at the same time. A 3D machine cuts in three directions at the same time. There are also four- or five-axis machines that cut/carve in three linear directions (x, y, and z), plus along one or two circular axes. Machines that are 2.5D and up are carvers as well as cutters.

FIGURE 13-3: From left to right, 2D, 2.5D, and 3D designs (photo courtesy of Paul Yanzick)

THE FUSION CAM WORKSPACE

Launch the emoji file and click CAM to enter the workspace (Figure 13-4). It looks like the Model workspace with a different menu (Figure 13-5). Let's go through the menu items. Most have elaborate submenus; hover over each entry for a pop-up description.

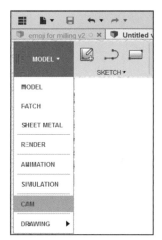

FIGURE 13-4: Click CAM to enter the workspace.

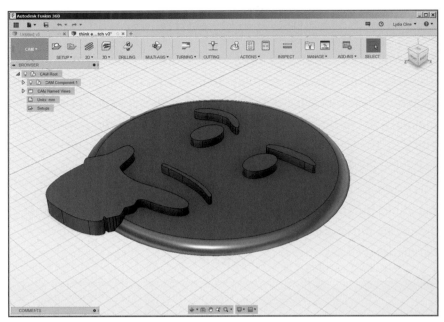

FIGURE 13-5: The CAM workspace

▶ **Setup:** Here you choose the kind of milling you want to do, orient the model, and specify the stock dimensions, where to place the model on the stock, and where the end mill will be relative to the material (Figure 13-6).

FIGURE 13-6: In Setup, choose milling type and general machining properties.

▶ **2D:** This menu has 2D operations, such as facing, contouring, and engraving. Figure 13-7 shows a tooltip for a facing operation, which quickly removes all material from the top of the stock to the top of the model, preparing it for machining.

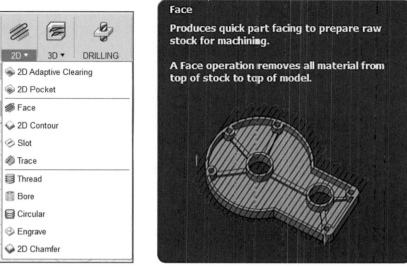

FIGURE 13-7: 2D facing is a 2D operation that removes all material from the top of the stock to the top of the model.

▶ **3D:** This menu has 3D operations such as adaptive clearing, pocket clearing, and scalloping (Figure 13-8). Such operations remove large quantities of material by making a series of z-layers through the part, and then remove them in stages from the top down.

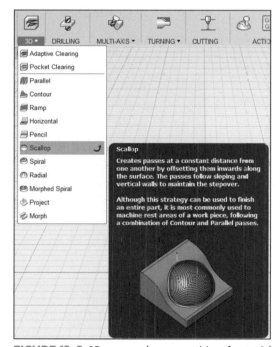

FIGURE 13-8: 3D removes large quantities of material.

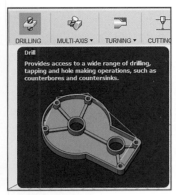

FIGURE 13-9: Make holes with the Drill menu.

▶ **Drill:** This menu contains drilling, tapping, and hole-making operations, such as counterbores and countersinks (Figure 13-9).

▶ **Multi-axis:** The options on this menu enable machining with the side or tip of the tool, or along selected surfaces (Figure 13-10).

▶ **Turning:** Use this menu for lathes, machines that cut parallel to the surface of a material being rotated (Figure 13-11). You can do both rough and finish toolpaths.

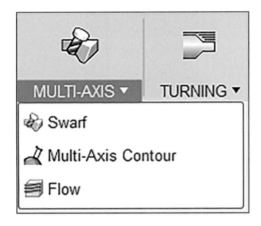

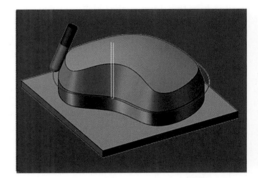

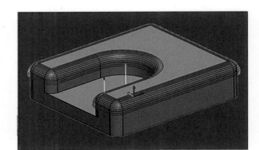

FIGURE 13-10: Multi-axis operations are possible.

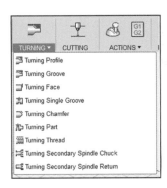

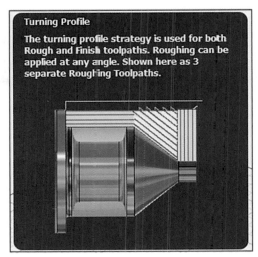

FIGURE 13-11: Turning is for lathes.

▶ **Cutting:** Cutting is a 2D operation used for waterjet, laser, and plasma cutters on flat stock (Figure 13-12).

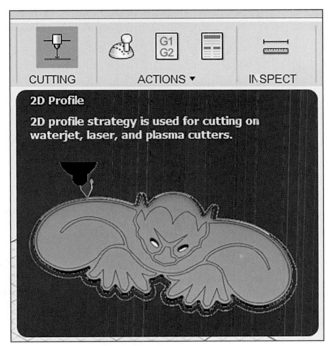

FIGURE 13-12: Cutting is a 2D operation.

▶ **Actions:** This menu has five options. We'll look at three: Simulate, Post Process, and Setup Sheet (Figure 13-13).

▸ **Simulate:** This option previews and simulates toolpaths and stock material removal. Controls include simulation speed and direction, tool, shaft, and tool holder visibility. Different cutting moves are shown with different colors. Playing the animation enables you to see if there are collisions (machine parts striking each other or the workpiece) and other problems.

▸ **Post Process:** This option converts the generic data into machine-specific code. There are customizable configurations for all popular CNC machines.

▸ **Setup Sheet:** This option generates an overview of the code—tool data, stock, workpiece position, and more—for the CNC operator to read. This sheet is also customizable.

FIGURE 13-13: Simulate, Post Process, and Setup Sheet are three options under Actions.

- **Inspect:** This menu lets you select an object and measure distance, angle, area and position. You can select by vertex, edge, face, body, or component (Figure 13-14).

- **Manage:** This menu provides Form Mill, Tool Library, and Task Manager options (Figure 13-15).

  - **Form Mill:** This option lets you make custom-shaped mills. They're created in the Tool Library from a sketch or a revolved solid body saved in a separate Fusion part file.

  - **Tool Library:** This option lets you manage tools for your individual projects. It has default tool libraries and you can import your own (Figure 13-16).

  - **Task Manager:** This is the interface for controlling toolpath generation. You can continue working inside Fusion 360 while generating toolpaths in the background.

- **Add-Ins:** This menu provides access to, and lets you run, scripts and add-ins (Figure 13-17). Create, edit, run, stop, debug, and manage scripts (apps that run and disappear when done) and add-ins (apps that run each time Fusion starts up). You can also access the Fusion App Store, which has free and for-pay apps.

FIGURE 13-14: The Inspect menu lets you measure the workpiece.

FIGURE 13-15: The Manage menu

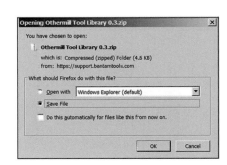

FIGURE 13-16: (Left) Default tool libraries; (right) a download library from the Bantam Tools website

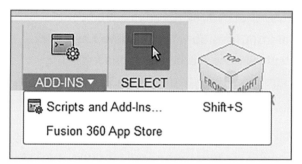

FIGURE 13-17: Scripts And Add-Ins

▸ **Select:** Use this option to deselect whatever tool you're in and activate the Select tool (Figure 13-18). From upper-left to lower-right is a selection window that selects everything inside it. From lower-right to upper-left is a crossing window, which selects everything it touches.

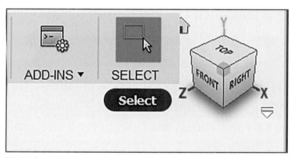

FIGURE 13-18: The Select tool

FIGURE 13-19: Return to the setup by right-clicking its Browser entry and choosing Edit.

GENERATE AN NC FILE FOR THE EMOJI PLAQUE

Let's use some of the functions just covered to make toolpaths and an NC file for the emoji plaque now. Note there's a Setups folder in the Browser. As you add operations, each will have an entry there. If you leave the setup before finishing, just click its Browser entry to return. If you want to edit a setup operation, right-click that operation in the Browser and choose Edit (Figure 13-19).

CREATE A SETUP

Choose Setup → New Setup (Figure 13-20). A yellow box representing a stock facsimile will enclose the model. By default, Fusion puts the model in the stock's center. Depending on the stock's size and the CAM techniques you choose, you may want to align the model to the surface of the stock or offset it. A triad will also appear. This is the World Coordinate System (WCS), and it appears as three arrows, with the origin at their intersection. The WCS arrow directions should match the positive direction of travel for the machine you're operating. You'll also see a dialog box, which has three tabs: Setup, Stock, and Post Process.

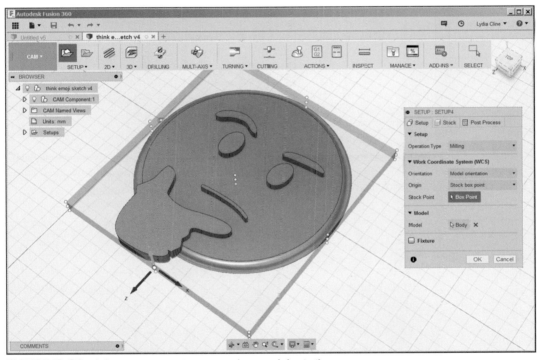

FIGURE 13-20: Click Setup → New Setup to make a stock facsimile.

On the Setup tab, note the Operation Type and Orientation options.

▶ **Operation Type:** You can choose from Milling, Turning, or Mill/ Turn And Cutting. Choose Milling, since it works for all types of milling machines.

▶ **Orientation:** Choose Select Z Axis/Plane & X Axis (Figure 13-21). Click the face of the model workpiece. The z arrow will orient straight up (if it points down, click the Flip Z Axis box). The green axis should point to the back of the machine, and the red axis should point to the right. By default, the triad is in the center and on top of the stock, but you can move it (Figure 13-22). Click the blue Box Point button to make circles appear on the workpiece. Then click any one of the circles to move the triad there.

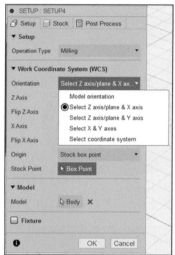

FIGURE 13-21: Orient the z-axis straight up.

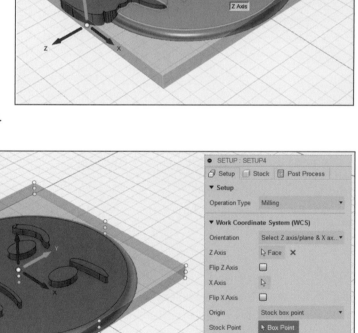

FIGURE 13-22: Use the Box Point button to relocate the triad.

On the dialog box's Stock tab you'll see Mode. Click the Relative Size Box option, which makes a stock that's larger than the model. Set the side offset to 10 mm so that there will be space to place clamps to hold the workpiece to the table and room for the cutting tool to move. The rest of the offsets can remain at their default, 0 (Figure 13-23). Alternatively, choose the Fixed Size Box option if you know your stock size.

FIGURE 13-23: Make the stock size a bit bigger than the model to provide space for clamps.

CREATE A 3D OPERATION

Click OK in the Setup box to close it. Then click 3D → Adaptive Clearing (Figure 13-24). Doing so clears large amounts of area. A dialog box with feed and speed options appears, which an experienced miller will finesse. Choosing feeds and speeds is an art. Too slow and you will burn the material; too fast and you will break the end mill. We'll leave the defaults. The URL for a wizard that calculates speeds and feeds is listed at the end of this chapter.

FIGURE 13-24: Choose the Adaptive Clearing option.

Click the Tool's Select button, and then click the New Tool icon. This brings up choices to customize a tool (Figure 13-25). We need a cutter that will fit between all the details of the workpiece. An experienced miller will make choices based on material and workpiece design. Select a holder out of the ones that are there (Figure 13-26).

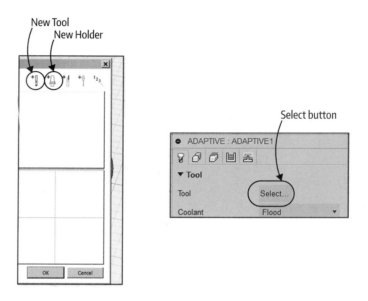

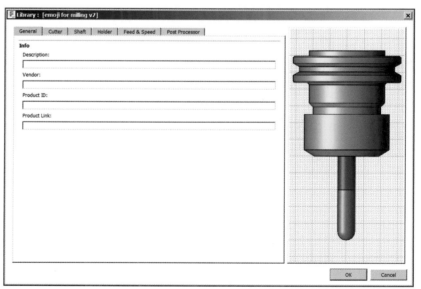

FIGURE 13-25: Click New Tool.

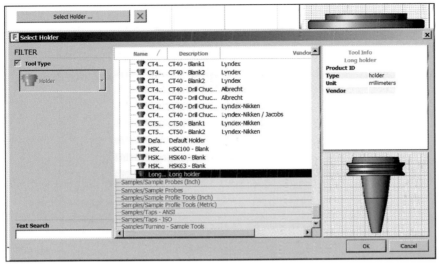

FIGURE 13-26: Select a holder.

The Select Tool dialog boxes let you specify Operation, Type, and Dimensions options (Figure 13-27). Here you can set the cutter's shoulder length, shaft diameter, flute length, body length, overall length, and diameter. For this design, a flat 10 mm end mill for the detail and a 10 mm ball mill for the rounded edge would probably work best. To edit a tool later, right-click it and choose Edit (Figure 13-28).

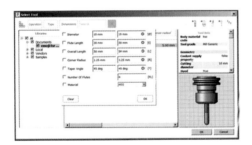

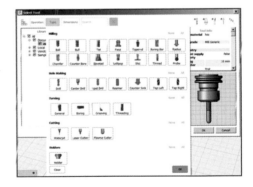

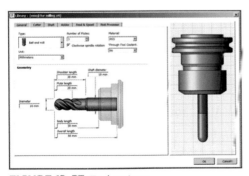

FIGURE 13-27: Tool options

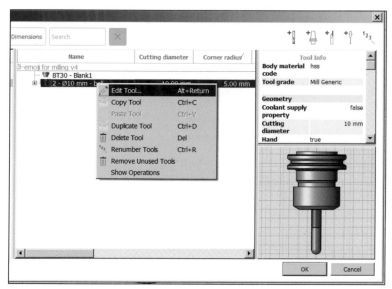

FIGURE 13-28: To edit a tool, right-click its entry.

Set Geometry, Heights, and Passes

There are five settings icons at the top of the Adaptive dialog box (Figure 13-29): Tool, Geometry, Height, Passes, and Linking.

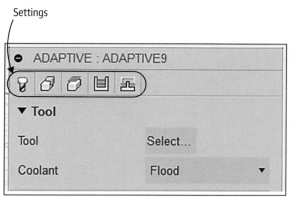

FIGURE 13-29: Icons at the top of the Adaptive dialog box

▸ **Tool:** Select a tool and specify your feeds and speeds.

▸ **Geometry:** Select what you want to mill.

▸ **Height:** Specify vertical dimensions of the toolpath.

▸ **Passes:** Configure the depth of passes, stepover (the distance the end mill moves when cutting), and stepdown (the z-direction distance the end mill is plunged into the material).

▸ **Linking:** Specify how the toolpath will begin and end.

We'll look at Geometry, Height, and Passes:

Geometry sets the machining boundary. Figure 13-30 shows the Silhouette option, with the bottom edge chosen as the path to cut. Select Stock Contours. If you deselect it, the whole piece of configured stock will be milled, including the offset we made earlier for clamps. Disable Rest Machining to disregard previous toolpaths (with Rest Machining, you can combine multiple toolpaths and do cleanup work, such as using a large end mill to do a lot of work and using a smaller one for fine cleanup/detail). Leave Tool Orientation and Model unselected.

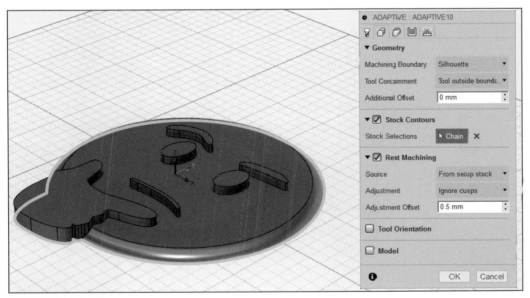

FIGURE 13-30: Silhouette

Height, or clearance, is the distance between the tool and the workpiece. The bit needs to clear the piece and any holders. It also needs to be long enough to go the needed depth. The default 15 mm clearance will work (Figure 13-31). Fusion automatically mills the whole area from the top of the stock to the bottom of the model. Unless you want to change this, leave the default settings on this tab.

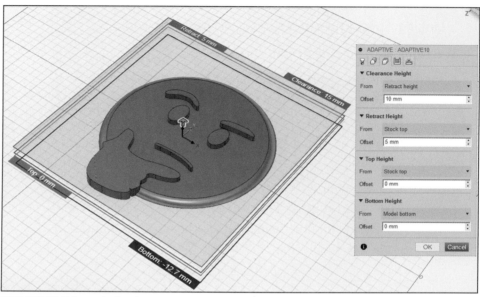

FIGURE 13-31: Set the heights.

FIGURE 13-32: The Passes options

Passes are the trips the end mill makes to cut out the material. A workpiece typically needs multiple passes. Figure 13-32 shows the Passes options. You can also use different techniques for passes; for instance, we might want to do two adaptive clearing passes for this workpiece, and one Contour pass (Figure 13-33) with a smaller end mill. Note the Optimal Load and Maximum Roughing Stepdown options. These set the amount of material the tool will cut on each pass. Set values for both that are generally no more than half the diameter of the tool.

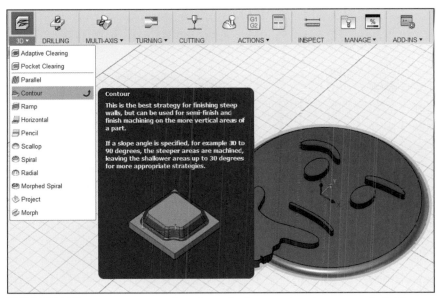

FIGURE 13-33: A contour pass and smaller end mill are good options to round off this workpiece.

SIMULATE

Once all the settings are entered, click OK. Fusion will generate the toolpath on top of the model. It may take a while; in the Browser you'll see a percentage of how much is done. Alternatively, on the Actions menu, choose Generate Toolpath. If a dialog box message appears saying that the selected operation is already valid, your toolpaths are good. You can click Yes to optionally regenerate them, or No to exit the dialog box. Then click Actions → Simulate (Figure 13-34). You'll see a facsimile of all the milling processes you set up (Figure 13-35) and buttons at the bottom to animate it. If all is well, move on to Post Processing.

FIGURE 13-34: Click Actions → Simulate.

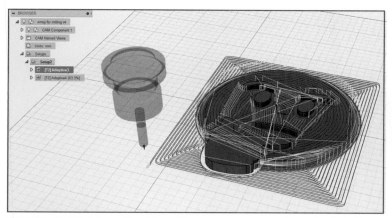 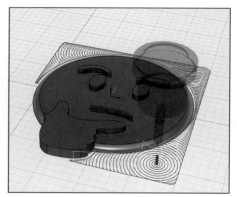

FIGURE 13-35: A facsimile of the milling processes shows the toolpath, indicating collisions and other problems.

POST PROCESSING

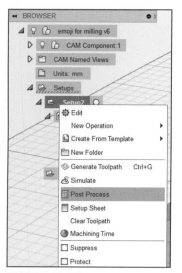

FIGURE 13-36: You can post-process the whole file or individual operations by choosing Post Process from the Browser entry.

Click Actions ➔ Post Process. This post-processes everything, creating a file that is machine or controller specific. If you just want to post-process an individual operation, right-click the Browser entry and choose Post Process (Figure 13-36).

Click Output Folder and choose your machine (Figure 13-37). Your NC file will generate (Figure 13-38), ready to download into your CNC machine.

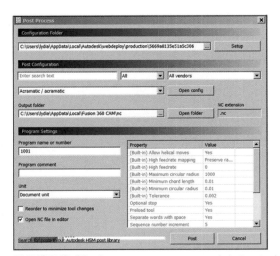

FIGURE 13-37: The Post Process menu and choices in the output folder

FIGURE 13-38: Part of the NC code generated

Additional Resources

Learn more about CNC: **http://en.wikipedia.org/wiki/Numerical_control**

Forum for ShopBot operators: **talkshopbot.com/forum/index.php**

Autodesk Fusion 360: Learn CAM **http://help.autodesk.com/view/ fusion360/ENU/?learn=cam**

Getting Started: Introduction to CAM and Toolpaths **www.autodesk.com/products/fusion-360/blog/ getting-started-introduction-to-cam-and-toolpaths/**

G-wizard, a speeds and feeds calculator for purchase: **www.cnccookbook.com/index/**

Index